202843

KT-425-138

Travel Journalism

Travel Journalism

Exploring Production, Impact and Culture

Edited by

Folker Hanusch
Queensland University of Technology, Australia

and

Elfriede Fürsich
Boston College, USA

Introduction, selection and editorial matter © Folker Hanusch and
Elfriede Fürsich 2014
Individual chapters © Contributors 2014

All rights reserved. No reproduction, copy or transmission of this
publication may be made without written permission.

No portion of this publication may be reproduced, copied or transmitted
save with written permission or in accordance with the provisions of the
Copyright, Designs and Patents Act 1988, or under the terms of any licence
permitting limited copying issued by the Copyright Licensing Agency,
Saffron House, 6–10 Kirby Street, London EC1N 8TS.

Any person who does any unauthorized act in relation to this publication
may be liable to criminal prosecution and civil claims for damages.

The authors have asserted their rights to be identified as the authors of this
work in accordance with the Copyright, Designs and Patents Act 1988.

First published 2014 by
PALGRAVE MACMILLAN

Palgrave Macmillan in the UK is an imprint of Macmillan Publishers Limited,
registered in England, company number 785998, of Houndmills, Basingstoke,
Hampshire RG21 6XS.

Palgrave Macmillan in the US is a division of St Martin's Press LLC,
175 Fifth Avenue, New York, NY 10010.

Palgrave Macmillan is the global academic imprint of the above companies
and has companies and representatives throughout the world.

Palgrave® and Macmillan® are registered trademarks in the United States,
the United Kingdom, Europe and other countries.

ISBN 978–1–137–32597–6

This book is printed on paper suitable for recycling and made from fully
managed and sustained forest sources. Logging, pulping and manufacturing
processes are expected to conform to the environmental regulations of the
country of origin.

A catalogue record for this book is available from the British Library.

Library of Congress Cataloging-in-Publication Data

Travel journalism : exploring production, impact and culture / [editors] Folker
Hanusch, Queensland University of Technology, Australia ; Elfriede Fürsich,
Boston College, USA.
pages cm
Includes bibliographical references.

ISBN 978–1–137–32597–6

1. Travel journalism.   2. Travel journalism—Study and teaching.   I. Hanusch,
Folker, 1975– editor. II. Fürsich, Elfriede, 1967– editori.

PN4784.T73T73 2014

070.4'4991—dc23   2014024793

Typeset by MPS Limited, Chennai, India.

Contents

List of Figures and Tables

Figures

Tables

UCB
202843

Acknowledgments

'In your face': Our book cover shows the production of one of the most typical visuals of tourist encounters, the close-up of an 'exotic' person. While we all have seen these types of photos in travel magazines, on this cover we experience the awkwardness and maybe absurdity of the encounter between photographer and 'model'. But it was especially the expression of the man that made us pause. A member of the Huli tribe in Papua New Guinea, he is part of a cultural performance staged for tourists. His expression is puzzling: proud, resigned, bewildered or even defiant. In one split second we see the complexity of the host and tourist/journalist interaction. The moment raises questions about authenticity, power, gender and technology that confound contemporary tourism and the work of travel journalists. It was exactly this complexity that triggered our interest in the topic of travel journalism.

This book brings an end to a journey that took shape almost three years ago, when we first discussed the slowly growing amount of travel journalism scholarship. Having both completed major projects on the topic, at times we had been frustrated with the fact that there was little academic work engaging with what we saw as an increasingly important area. Yet recent publications had given us hope that the field was approaching a critical mass. The idea was thus born to take stock of where we were, and to explore the way forward. The response to our ideas was tremendous, with a number of emerging and established scholars accepting our invitation to contribute chapters. We would like to thank them all for their dedication to this project, the insights they shared with us and their openness to our feedback.

We would also like to thank Palgrave Macmillan's commissioning editor, Felicity Plester, who took up our idea with much enthusiasm. She was a great supporter of the project from the moment we proposed it. We also appreciate the thorough work of the editing and production team at Palgrave Macmillan, especially Chris Penfold who expertly guided the book through the preparation and production phases. The two external reviewers also provided much-appreciated comments.

In addition, Folker would like to thank the University of the Sunshine Coast, which funded a number of his earlier projects on travel journalism, thus enabling him to explore a field that captivated him ever since working as a travel agent after high school. Thanks are also due to the many colleagues who gave him feedback on his work as reviewers of journal submissions and at conferences. Finally, Folker owes gratitude to his wife Stephanie, for providing the original inspiration to conduct studies in this area. After he had been investigating news representations of death for a number of years, it was she who urged him to explore something a bit more positive and 'fun', like travel journalism. He is glad he took up the challenge.

Elfriede would like to thank John Hartley, the editor of *International Journal of Cultural Studies*, who published her first article on travel journalism. She is also indebted to her student assistants at Boston College who helped with the research over the years. Moreover, the continued support of Boston College and current and former colleagues at the Communication Department, in particular chair Lisa Cuklanz, has been instrumental. Her husband CB Bhattacharya always seems to be planning a trip to somewhere and keeps the family going with his question 'Where should we go next?' Her gratitude goes to him and their son Felix. Both will always be her favorite travel companions.

Much work remains to be done in the ongoing development of the field of travel journalism, particularly as the practice and profession grows further around the globe. We hope that this volume provides a helpful update on the state of the field, generating ideas and inspiration for future work.

Notes on Contributors

Jiannu Bao currently teaches at the Department of International Journalism and Communication, School of English and International Studies, Beijing Foreign Studies University, China. Her research interests include the comparative study of lifestyle journalism, especially travel journalism, and environmental journalism. She worked for the Chinese state Xinhua News Agency from 1992 to 2001, first as a copy editor in the newsroom of the Department of Home News for Overseas Services and later as a staff writer for the China Features and Services, writing mainly for overseas publications including a tourism trade magazine.

Christine N. Buzinde is Associate Professor in the School of Community Resources and Development at Arizona State University, Phoenix, Arizona. Her research focuses on two areas: community development through tourism and the politics of tourism representations. Her work on representations regards tourism texts as cultural repositories through which inclusion/exclusion, North/South and core/periphery can be understood. She examines texts as sites wherein entanglements of power and oppression as well as depictions of agency and resistance can be unveiled. Dr. Buzinde's work on development adopts a grassroots approach, and aims to understand the relationship, or lack thereof, between community well-being and tourism development within marginalized communities. She conducts research in Tanzania, Mexico, India, Canada and the US. She has published numerous articles in tourism, geographical and cultural studies journals.

Divya Choudary holds an MA in Communication and Print and New Media Journalism from the University of Hyderabad in India. She is now assistant editor at *Teacher Plus* and is also the editor of the English section of the *Teachers of India – Azim Premji* Foundation portal. Her interest lies in tying together narratives and technology to create different ways of storytelling.

Ben Cocking is Principal Lecturer in Media and Cultural Studies at the University of Roehampton, London, UK. He teaches on the Journalism and Media & Culture programs at undergraduate and postgraduate levels, including PhD supervision. His research interests include travel journalism, representations of the Middle East in news media and Arabist and British post war travel writing. He has recently published articles in *Jomec, Journeys, Journalism Studies* and *Studies in Travel Writing*.

Brian Creech is Assistant Professor of Journalism at Temple University in Philadelphia, Pennsylvania. His research interests include cultural studies of technology, post-structural and post-colonial theories on media, journalistic practice as a form of public discourse, and the relationship between media, war and terrorism. His current research project investigates the ways that new media technologies interact with traditional journalistic practices to foster or limit new forms of political subjectivity and democratic communication.

Maja Sonne Damkjaer is a PhD student in the Department of Aesthetics and Communication, Media Science at Aarhus University, Denmark. Her project aims to elucidate the mediatization of parenthood through an exploration of the main ways digital media have become intertwined with the initial, formative phase of parenthood. She has an MA in Media and Communication from Aarhus University and a Bachelor of Education from Silkeborg Seminarium. Since her MA she has been working as a research assistant and lecturer, primarily at Aarhus University, teaching media text analysis, key media theory and journalism genres. Lately she has been teaching strategic communication and concept development at the Danish School of Media and Journalism in Aarhus. Maja's interest is media sociology, mediatization and media aesthetics and the link between the three, especially in relation to lifestyle journalism, tourism and parenting. She focuses on visual culture and studies television, films, magazines and digital media.

Andrew Duffy is Assistant Professor at the Wee Kim Wee School of Communication and Information at Nanyang Technological University in Singapore. His research interests include the use of online democratic information sources and user-generated content in travel planning, the influence of soft journalism and the changing face of journalism practice in an age of technological change. Previously he was a journalist on magazines and newspapers as a writer and editor in the UK and Singapore for 15 years. He worked as a travel writer and editor for five years, before moving into contract publishing for clients including Thomas Cook and the British Tourism Authority.

Elfriede Fürsich is Research Associate Professor of Communication at Boston College. She specializes in issues of globalization, mobility, media culture and criticism. She has investigated the discourse on globalization in media ranging from travel programs or music reviews to business journalism and African Internet sites. She also wrote an expert report on 'Media and the Representations of Others' for the UNESCO World Report 'Investing in Cultural Diversity and Intercultural Dialogue' (2010). In 2005, she was Visiting Fellow at the University of Hyderabad, India. Currently, she is Visiting Professor at the Erasmus Mundus Master's Program 'Journalism, Media and Globalization' at the University of Hamburg, Germany.

Folker Hanusch is Senior Research Fellow in Journalism at Queensland University of Technology. His main research interests are in lifestyle and travel journalism, comparative journalism studies, journalism culture, indigenous journalism and journalism and suffering. His work has appeared in international journals such as *Media, Culture & Society, International Communication Gazette, Journalism, Journalism Studies, European Journal of Communication* and *Journalism and Mass Communication Quarterly*. He has also published an edited collection on lifestyle journalism, as well as authored books on journalism across cultures and news representations of death and dying.

Cathy H. C. Hsu is a professor in the School of Hotel & Tourism Management, The Hong Kong Polytechnic University. Her research foci have been tourism destination marketing, tourist behaviors, hotel branding, service quality and the economic and social impacts of casino gaming. She has authored several textbooks on tourism marketing and published over 200 refereed articles. She is the recipient of the John Wiley & Sons Lifetime Research Achievement Award in 2009.

Lyn McGaurr is a research associate at the University of Tasmania, where she completed PhD in Journalism, Media and Communications. She began her journalism career in 1979 as a researcher and trainee reporter at the Hobart, Tasmania, office of Australia's public broadcaster, the ABC, during a period of intense environmental conflict over proposals to dam the Franklin River in the state's southwest. Later, she developed a lasting interest in climate change while briefly responsible for media relations for an Antarctic scientific research organization in the lead-up to the 1997 Kyoto negotiations. She has worked for Lonely Planet as a travel guide editor, and, in the late 1990s, was the updating author of the second edition of its *Tasmania* guide. She has been a senior book editor for Penguin Books Australia, and has also worked for Tourism Tasmania from 2002 to 2008. Her journalism honours thesis surveyed coverage of climate change and nuclear energy in the *Australian* newspaper before and after the release of the 2007 IPCC report, while her PhD thesis considered international travel journalism about Tasmania from a cosmopolitan perspective. She is currently writing a book about travel journalism and environmental communication.

Steve Pan is Assistant Professor in the School of Hotel & Tourism Management, The Hong Kong Polytechnic University. He has more than 10 years of experience working in journalism. His research focuses on the representation of a destination in travel journalism, including print and broadcast media and tourism TV commercials. He has published articles in journals such as *Cornell Hospitality Quarterly, Journal of Hospitality and Tourism Research, Journal of Travel Research, Journal of Travel and Tourism Marketing,* and *Tourism Management.*

C. Bjørn Peterson is a PhD student in the School of Community Resources and Development at Arizona State University. His research focuses on community-based development, inclusive process design, development-related narratives and philosophy of development.

Bryan Pirolli is an American PhD candidate and instructor at the Sorbonne Nouvelle in Paris, where he is exploring travel journalism's evolution in a digital world. He investigates how information sources such as *TripAdvisor,* blogs and other user-generated content are challenging traditional journalistic standards. He is a blogger (www.bryanpirolli.com) and has also worked as Paris correspondent for CBS's *SmartPlanet, CNN Travel, Time Out, Fodor's, Zagat* and *DK Eyewitness* travel guides.

Usha Raman is an associate professor in the Department of Communication, University of Hyderabad, India. A writer, health communicator and non-profit advocate, she is also Editor of *Teacher Plus,* a monthly magazine for schoolteachers in India. Her research interests include cultural studies of science, science and health communication, children's media and the impact of new media on individual and community life.

Wiebke Schoon is Research Associate and Instructor at the Institute for Journalism and Communication Studies and PhD candidate at the Graduate School of Media and Communication, University of Hamburg. Her current research interests lie in the field of transcultural media and communications studies. She is especially interested in journalism's role as a mediator and actor in processes like globalization and Europeanization. In her dissertation she combines the sociological concept of cosmopolitanism with a cultural understanding of journalism to investigate how German travel journalism has discursively constructed European and other regional spheres over time.

Anne Marit Waade is Associate Professor of Media Studies, Aarhus University, Denmark. Her research is within media aesthetics, visual culture, mediated tourism and branding culture. She has recently published articles on crime fiction series and crime tourism (www.krimiforsk.aau.dk), travel series and travel journalism (www.tvunderholdning.au.dk), Nordic Noir (2013) and the books *Wallanderland* (2013), about the Scandinavian crime series production, film tourism and regional development, and *Medier og turisme* (2009) together with Jakob Linaa Jensen. She has co-edited several volumes and journals; *Re-Investing Authenticity* (2010) with Britta Timm Knudsen, *Skandinavisk Krimi* (2010) with Gunhild Agger, *Northern Lights* special issue on *Crime and Media* (2011) with Gunhild Agger.

Eunice Eunjung Yoo is Lecturer in Hospitality and Tourism in the School of Hospitality and Tourism Management at the University of Surrey, UK. Her research centers on socio cultural aspects of tourism and hospitality, focusing on critical approaches to tourism representations and destination imagery within the context of various forms of media. In particular, she explores the role of food and its cultural and symbolic significance through various forms of travel media, where food, culture and tourism overlap.

1
On the Relevance of Travel Journalism: An Introduction

Folker Hanusch and Elfriede Fürsich

Not many people will be instantly familiar with British woman Dale Sheppard-Floyd, but – at least symbolically – she represents a significant milestone in the development of travel and tourism. In fact, the milestone was so significant that the United Nations World Tourism Organization booked Madrid's venerable Museo del Prado to announce to the world's media her visit to Spain on 13 December 2012. For Ms Sheppard-Floyd's arrival for a three-day trip meant that more than one billion times in that year, someone had crossed a border as a tourist. An astounding number, considering that, in 1950, there had been only 25 million tourist arrivals worldwide, and even only two decades previously – in 1990 – the number had been less than half at 435 million arrivals (World Tourism Organization, 2012a, 2012b). While people have traveled for pleasure for millennia (Towner, 1995), tourism really came into its own with the expansion of the middle classes in the 19th and 20th century, and today it is considered the world's largest business sector, with unprecedented numbers of people venturing outside of their immediate environments to explore the world around them. In 2012, travel and tourism's total contribution to the world economy amounted to a staggering $6.6 trillion, or 9 per cent of GDP (World Travel & Tourism Council, 2013). More than 260 million jobs were generated by it worldwide, which equates to one in every 11 jobs across the globe. While there were some hiccups during the Global Financial Crisis, growth in 2012 was stronger than in other industries, such as manufacturing, financial services and retail (World Travel & Tourism Council, 2013).

The phenomenal rise of tourism, particularly during the past 50 years, has led to a concurrent interest from scholars, in particular in the growing number of tourism and travel studies programs at universities. Media and journalism play an immensely important role in communicating destinations to potential tourists. Aside from friends' recommendations, much of the information that travelers receive is through general media coverage, as well as more specific travel information in print, broadcast, or, increasingly, online. Yet, the role that media and journalism play in ascribing meaning to tourism and producing tourist destinations has only recently garnered the attention from scholars around the world. One reason for this lies in media scholars' traditional inclination to focus on hard news journalism, rather than its softer varieties. This is slightly curious, as many commentators have identified a shift in journalistic output towards these softer types, in particular the area of lifestyle journalism, which has experienced a drastic rise over the past few decades (Hanusch, 2012a). International news reporting, in particular, appears to have experienced a decline in its authority and the amount reported by mainstream media, which have corresponded with a rise of non-fiction entertainment such as travel content as a global media genre (Fürsich, 2003).

Invariably, travel journalism has received 'flak' from the broader journalism and communication research community for it is clearly a market-driven type of journalism. Thus, travel journalism is still widely regarded as a 'frivolous topic' for research, even more than ten years after Fürsich and Kavoori (2001) noted this positioning towards the field. The general view is often that anyone can go on holidays, and anyone can write about them. As the US travel journalist Thomas Swick (1997, p. 424) has pointed out, 'of the special-section editors at a newspaper – travel, fashion, food, home and garden – only [travel] occupied a position that is viewed as requiring no particular expertise'. Swick also argues the field has been seen as one in which anyone can work: 'Not only do most people travel, most people write postcards when they do: ergo, most anyone can be a travel editor' (1997, p. 424). And this lack of esteem continues despite travel information in the media being cited as an increasingly important source of information for tourists, at least in the early stages when tourists form their motivation for visiting a destination (Nielsen, 2001, pp. 126–29).

The general reluctance among journalism and media researchers to expand into the softer types of journalism has slowly given way to a sense of acceptance, however. Over the past decade, research in travel journalism has reached a critical mass, with a number of scholars engaging with the production, content and reception of travel journalism and travel media. A similar trend has taken place in other fields.

Anthropology, cultural geography, sociology and sociolinguistics, along with the emerging field of critical tourism studies, have interrogated the social, cultural and discursive dimensions of tourism practice (see for example, Burns and Novelli, 2009; Graburn and Barthel-Bouchier, 2001; Jaworski and Pritchard, 2005; Scott and Selwyn, 2010; Thurlow and Jaworski, 2010; Urry and Larsen, 2011). It is precisely the significance of leisure in contemporary society that makes the study of tourism and of travel journalism such a fertile field for research.

This growth in academic attention to the field was the catalyst for the production of this book. Having followed the development of the field over the past 10 to 15 years, we decided that it was time to take stock of what we know about travel journalism, critically examine our existing approaches and open up avenues for future research. For the first time, then, this collection will provide a comprehensive introduction to the field of travel journalism studies.

The amount and variety of travel and tourism as media content has grown along with the rise of digital technologies, the increasing commercialization of media output and the fragmentation of audiences. Besides specialized travel magazines, newspapers produce ever-larger travel sections, and entire cable and satellite television channels are devoted to travel and tourism. Moreover, the Internet now plays a central role in the media discourse on travel. This volume explores a wide variety of media and types of journalism from a range of cultural contexts, and pays special attention to the recent developments in professional practice.

Significance of travel journalism

A number of reasons support the argument for a closer scholarly engagement with travel journalism.

Boom of the tourism industry

As mentioned above, tourism has become a significant global economic force. Whereas most traveling still takes place within the borders of the home nation, for an increasing number of people international travel is no longer exceptional. The statistics on international arrivals and receipts mirror that development. In 1956, when the World Travel Organization started publishing statistics, worldwide international arrivals were estimated to be 50 million, with international tourism receipts of about $4 billion. Within 40 years, in 1996, arrivals had increased to 594 million with receipts of $423 billion (Waters, 1997). By 2012, these numbers have risen even more to 1.04 billion international travelers and $1.08 trillion in receipts (World Tourism Organization, 2013).

Whereas the United States, Spain and France are still the top earners in the international tourist business, China now ranks as number 4 of international receipts. Overall, the traditional tourism countries are losing market share to periphery countries and emerging economies. For example, the strongest growth rates were experienced in Southeast Asia (plus 10 per cent) and Eastern Europe (plus 7 per cent). The sector is projected to expand at 4 per cent annually in the long-term (World Tourism Organization, 2013).

Tourism as an instrument of economic policy

Countries rely to a different degree on tourism as a major employer and source of Gross National Product. Many developing countries and emerging economies hope to make tourism – as a presumably 'clean industry' – one of the main sectors of their economy. But Western countries such as Austria or Spain also depend on tourism as important revenue and employment sources for their country's economy.

Overall, increasing automatization in Western countries and the service-oriented information economies have steadily decreased the average weekly and annual working hours during the past 50 years. This has led to the development of leisure societies where tourism is no longer an activity of elites (e.g., Urry, 1990). Moreover, tourism is no longer practiced only within the Global North or from 'the West to the rest'. The end of the Cold War changed the restrictions on travel in many former communist countries and created a whole new patronage of travelers. The economic success of many emerging economies has expanded the upper middle class in these countries, resulting in more people who value the experiences of foreign travel and have the extra money to spend on travel. For example, China has been the largest travel market for outbound travel since 2012 (World Tourism Organization, 2014). Thus, the changing geopolitical and economic situation in many parts of the world has brought new constituencies to an already booming travel industry.

In addition, during the past decade tourism development has become a much-championed instrument used by economic policy makers across the world to push for a creative economy and an anticipated rise in jobs in the so-called 'creative class' (Florida, 2012). As critics (e.g., Flew, 2012, pp. 159–82; Ross, 2009, pp. 15–52) explain, cities and regions across the world, especially in the post-industrial Global North, compete against each other in an intensive struggle for investments, innovation and sustainable economic development. To attract much-sought after talent, political leaders use tourism strategies (including the gentrification of de-industrialized spaces or events marketing) to create a positive image of their cities as destinations to visit and to live in.

Travel journalism as an important site for international communication research

Paralleling the growth of tourism as a global industry has been the exponential growth of travel journalism. In addition to the traditional travel section in most major national and regional dailies, a large number of general travel magazines are published, along with a prodigious number of specialized travel publications dealing with interests as diverse as rock climbing or cruise vacations. The broadcast media offer specialized travel programs and celebrity travel shows, and a number of countries have entire cable channels devoted to the subject. The Internet is another highly successful outlet for travel-related information. Travel sites of online services and travel-related webpages are among the most accessed websites on the Internet. An online survey by the travel website TripAdvisor (2013) shows just how important travel review websites have become. Around two-thirds (69 per cent) of travelers used these sites in their planning, while only 30 per cent used magazines and brochures. Further, 93 per cent of respondents said their booking decisions had been impacted by online reviews, and 51 per cent said they had written a review of an accommodation themselves after a trip.

The expansion of international tourism has affected the media industry in two ways. First, affluent groups from an increasing number of countries are traveling for pleasure or business. This development generates audience interest for travel-related journalism and information as a media topic. Potential travelers will be interested in this kind of journalism for advice and entertainment. Moreover, the growing global 'middle-class' will understand travel, especially international travel, as a desirable private status goal while using mass-mediated travel as *ersatz* experience as long as they cannot afford actual trips. Second, the tourist industry has generated a larger market for travel advertising and public relations, looking especially for media outlets that promise a targeted and receptive audience. These interrelated trends are exemplified in this book by Jiannu Bao (Chapter 8) who studied the exponential growth of the Chinese travel media market. These developments continue to stimulate a growing market for specialized travel journalism on a global scale.

The increased prominence of travel journalism has relevance for media scholars. This is especially evident when we consider how travel journalism functions much like international news to provide both information and cultural frames for 'others'. International communication research has traditionally focused either on the spread of news and entertainment or on advertising in a global market (e.g., Reeves, 1993). When looking at the way national media represent foreigners and foreign cultures, studies tend to analyze international news content in newspapers or on television. However, audience interest in 'hard' international news is waning while media representations of 'others' remain decisive factors in this era of globalization. Therefore, a research agenda of international communication studies can gain from evaluating other media genres. Examining travel journalism is an important strategy for analyzing the dynamics of globalization. Thus, instead of criticizing travel journalism as trivial cultural celebration we can ask what discourse is created within media representations of travel. This approach allows us to interrogate the cultural and ideological assumptions upon which such constructs are based.

Defining travel journalism

In order to frame the discussions in this book, as well as in the wider field, it is, of course, important to define what exactly we mean by the term 'travel journalism'. Often, the terms 'travel writing' or 'travel literature' on the one hand, and 'travel journalism' on the other, are used interchangeably, leading to problems of differentiating between the two. It is important to draw a distinction, however, mainly because the term 'journalism' for most people invokes certain norms and ideals. And indeed there are often different standards, in that travel writing more generally allows the inclusion of fictional elements and other literary license that would not be accepted in traditional news media. This has implications for research approaches, too. The literary studies approach often employed for the analysis of semi-fictional accounts of travel, for example by Paul Theroux or Bill Bryson, cannot convincingly explain the unique economic and public situation of journalistic work. Instead, it is travel journalism's position bound to professional ideas of journalism in its representation of distant places and people that makes it such a distinctive site for research. An additional complication is that quite regularly travel writers alternate between producing travel books and writing for other print or online media, blurring the boundaries for audiences.

Perhaps it is important first to clarify what we mean by the term journalism. Zelizer (2004, p. 3) highlights five definitional sets through which journalism can viewed: 'as professionalism, as an institution, as a text, as people, and as a set of practices'. Schudson (2003, p. 11) adopts a functional definition and writes that 'journalism is the business or practice of producing and disseminating information about contemporary affairs of general public interest and importance'. While he warns about normative definitions of journalism, and acknowledges that there are other types of journalistic practice, Schudson nevertheless focuses on journalism that relates to political affairs, because it is 'that part of journalism that makes the strongest claim to public importance' (2003, p. 15). As we have pointed out, this relatively narrow focus on journalism's relationship with politics surrounds much of the existing academic work in the discipline. As a result, journalism that occurs outside the normative ideal has 'become denigrated, relativized, and reduced in value alongside aspirations for something better' (Zelizer, 2011, p. 9).

However, despite the abundance of normative definitions of journalism, there are also more straightforward explanations. For example, McNair (1998, p. 9) sees journalism as 'an account of the existing real world as appropriated by the journalist and processed in accordance with the particular requirements of the journalistic medium through which it will be disseminated to some section of the public'. Such an inclusive definition does not judge or privilege one kind over the other. The emphasis on the 'existing real world' is an important marker when referring to journalism, in order to differentiate it from fictional accounts.

Thus, we argue that while some travel writing can be regarded as travel journalism, the latter is more closely connected to the professional notions around fact, accuracy, truthfulness and ethical conduct of journalism. The important criterion of distinction here is in Fürsich and Kavoori's (2001) original definition, based on Hartley's (1996) notion of journalism as a textual system:

> The most important *textual* feature of journalism is the fact that it counts as true. The most important component of its *system* is the creation of readers as publics, and the connection of these readerships to other systems, such as those of politics, economics, and social control. (Hartley, 1996, p. 35, emphasis in original)

The notion of journalism's truth claim is central to differentiating between travel journalism – bound by journalistic notions of reporting on the real – and travel writing, which may include fictional elements. Further, it is important that the definition is able to accommodate work for any medium, be it newspaper travel sections, travel magazines, television travel shows or travel websites. Travel writing, by virtue of its very use of the term 'writing' itself, is still bound to the written form in contrast to the increasingly common multimedia practice of travel journalism. In fact, as some of the chapters in this book note, travel journalism is increasingly being conducted online, linking new creators and publics to its output.

The special exigencies of travel journalism mean that travel journalists are likely to be a much more heterogeneous group than news journalists. For instance, there is a very large component of freelancers who work in the industry, many of whom are organized in professional societies such as the Society of American Travel Writers, the North American Travel Journalists Association or the British Guild of Travel Writers. In addition, most media rely on regular news journalists and freelancers to contribute travel stories. Those who produce travel journalism are thus not always 'experts' in the field, which may have some impact on the content they produce.

The biggest challenge for travel journalism as a profession is the increasing number of amateur writers who generate travel information online. This is a similar situation to the one that mainstream news journalism finds itself in (Bruns, 2005). As several chapters in this book highlight, there is a wide spectrum of engagement from non-professional writers in travel journalism. While some input is limited to a few lines of review on TripAdvisor or a few pictures on Flickr, other work is produced by influential travel bloggers who often create written and visual content for various traditional media as well. This development has led to a paradoxical situation: While new technologies have made it easier than ever for someone to enter the field and produce travel content for a large audience, the abundance of voices has made it more difficult than ever to actually make a living as a travel journalist. While it may be tempting to dissociate all types of amateur or citizen journalism on travel as a poor copy of 'real' travel journalism, the actual variety, reach, and impact of these offers warrants the inclusion of amateur efforts in definitions of journalism. Whereas a professional journalism background, or the ability to make a living as a journalist are no longer discerning factors, in this book we do concentrate on journalistic work that is done with the expectation of income at a reasonable time. This decision allows us to structure the field and separate from it shorter contributions by users who post only once in a while.

Dimensions of travel journalism

When examining the existing studies on travel journalism we can discern four important aspects, or dimensions, that can further help to define the term, and the field more broadly.

The representation of foreign cultures

The most dominant concern of scholars studying travel journalism so far has been its role in representing other cultures and nations. The main purpose of travel journalism is to represent the Other. Based in the cultural studies tradition (e.g., Fürsich, 2002) and critical tourism studies (e.g., Santos, 2004), these studies have focused predominantly on the content of travel journalism. Often these studies have demonstrated that travel journalism presents a friendly and celebratory, albeit exoticizing and stereotypical discourse of the Other. Moreover, from studies in the field of marketing in tourism we do know that news media reports can influence the images that tourists have of a destination. Beerli and Martín (2004), for example, found that organic (such as friends) and autonomous sources (such as independent media reports) significantly influenced some aspects of the destination image of tourists in Lanzarote. In a survey of tourists' images of Tibet, Mercille (2005) also discovered a relatively strong influence of mass media images on what tourists expected when they visited the country for the first time.

The ethics of travel journalism

A second dimension that has attracted much discussion relates to travel journalism's tacit allegiance to both advertising and the travel industry. In fact, travel journalism is a highly charged discourse beleaguered by public relations efforts of the private travel industry and by government-sponsored tourism departments (Hanusch, 2012c). In addition to the public relations saturation, travel journalism often exists in symbiotic relationship with advertising. Travel is mostly covered in special sections of newspapers, in magazines or on television shows, which almost exclusively find their advertisers within the travel industry itself. If anything, the online world of travel information has even intensified the collaborations between the tourism industry and professional and amateur journalists as new forms of sponsorship and linking take hold. All these practices place many travel journalists in a difficult position between major interest groups.

Traveling to, and reporting from, distant places is an expensive exercise, and most news organizations are unable to pay for all the travel experiences about which they publish stories. This is even more the case in the current economically precarious environment for news media, which has led to the decrease in foreign news reporting in the first place. Few academic studies have investigated the relationship between travel industry and journalists systematically – even though it has been a dominant talking point in practitioner reports about the constraints of travel journalism. The professional attitude is also not always as stringent as in news departments. Most travel journalists realize that free travel or accommodation is necessary for them to do their job, but they believe their editorial output is not necessarily influenced by this as much as some might imagine (Hanusch, 2012b, 2012c). US travel journalist Elizabeth Austin confirms that 'the writers of most junket-based pieces generally sing the praises of their hosts' accommodations, let's face it: Travel publications celebrate travel' (1999, p. 10). Yet she argues that stories that have been paid for by the publishers themselves may even be more biased. After all, the publisher wants an outcome for their expense – potentially leading an author to portray a destination or experience more positively than it was, making it a better story in order to justify the trip. Nevertheless, the issue of disclosure is often debated. Interestingly, an analysis of newspaper travel sections has found that articles which carried a disclosure note actually contained more in-text advertising (i.e. overly positive coverage of a travel provider) than those which did not (Hill-James, 2006).

Travel journalism's market and consumer orientation

A third area that differentiates travel journalism from most hard news journalism is its market orientation. As a type of lifestyle journalism most travel journalism considers audiences unashamedly as consumers, rather than citizens, even though that does not mean some travel journalists do not also try to be critical in their reporting. Hanitzsch (2007) differentiates between journalism in the public interest and journalism that addresses audiences as consumers, which gives them 'what they want' and places high emphasis on entertainment. Typical travel journalism culture would be expected to be ranked as high market orientation according to Hanitzsch's (2007, p. 375) definitions, as its aim is the 'blending of information with advice and guidance as well as with entertainment and relaxation'. Indeed, this was found in a study of Australian travel journalists, which noted that they primarily identified with roles that relate to the discovery of new and unique travel experiences and an aim to provide useful, interesting and entertaining information (Hanusch, 2012b). They thus subscribe to a service function, in line with a traditional understanding of lifestyle journalism that provides 'news you can use', as well as a commercial orientation in the vein of 'soft news'. An interesting aspect in this regard is that most newspapers publish genuine news about the tourism industry in their business sections, where stories are produced by dedicated business journalists, not travel journalists – a distinction which may allow for a more narrow definition of what constitutes a travel journalist.

Motivational aspects of travel journalism

Another dimension of travel journalism is the way in which travel journalists are motivated, that is, how they engage with their audiences as prospective travelers. Fürsich (2002) differentiates between three types of tourism coverage, ranging from uncritical celebrations of travel to critical perspectives on actual trips to reports that problematize tourism and the industry more generally. By engaging with tourism in these different ways, the media help construct differing ideal types of tourists. Such work is grounded in Urry's (1990, 1995) seminal work on the sociology of tourism, which examines the multitude of tourists' motivations for and behavior during travel. Urry popularized the notion of post-tourists, who experience a multitude of meanings on their trips, rather than existing structural typologies developed by tourism researchers such as Cohen (1979), Plog (2001), or Smith (1989; Smith and Brent, 2001). Motivation is an important dimension, because it allows us to link concepts in journalism scholarship on what constitutes journalism with tourism studies' notions of what tourists expect.

Following this brief review of what we consider the four key dimensions of travel journalism, coupled with the earlier explication of definitions of journalism and what separates travel journalism from travel writing, we can now arrive at what we hope to be a reasonably inclusive definition. Hence, we define travel journalism *as factual accounts that address audiences as consumers of travel or tourism experiences, by providing information and entertainment, but also critical perspectives. Travel journalism operates within the broader ethical framework of professional journalism, but with specific constraints brought on by the economic environment of its production.*

Overview of chapters

Split into four parts, this collection presents work from well-established scholars in the field along with emerging authors in an effort to promote and illustrate the relevance of studying travel journalism.

Covering cultural, social, political and media-related dimensions, the book is interdisciplinary in nature, combining journalism, communication and cultural-critical media studies approaches with tourism and globalization studies and cross-cultural communication approaches. Authors and topics originate from North America, Africa, Europe, Asia and Australia, providing for a truly international perspective on many of the issues concerning travel journalism today.

The 13 chapters that follow this introduction provide a critical discussion of theoretical approaches, in-depth studies on travel journalists, content and impact, as well as ways in which travel journalism can be understood through the lenses of postcolonialism, sustainability and cosmopolitanism. Using qualitative and quantitative methodologies, the contributors deal with a wide range of travel journalistic media, including newspapers, magazines, television and online publications. They identify important trends in, and challenges for, travel journalism research in the intermediate future.

The chapters in this book are organized into four parts, the first of which revolves around strategies that scholars may adopt when studying travel journalism. The aim here is to outline themes along which research may be conducted, as well as to provide some useful categorizations for future research. In Chapter 2, Elfriede Fürsich and Anandam Kavoori update and extend an article, published in 2001, which has become a seminal piece of work for the field as a whole. They retrace their critical framework for studying travel journalism, which is structured by issues of periodization, power and experience, and they assess the progress that has been made in these areas over the past decade. They also introduce a new issue around the notion of mobility, based on recent research in this field. Chapter 3, contributed by Maja Sonne Damkjaer and Anne Marit Waade, provides readers with a typology for characterizing television travel shows. Their analysis of travel series broadcast on Danish television channels from 1988 to 2005 results in a typology of ten different sub-genres, which will be helpful to guide future studies in this field. Chapter 4 combines approaches from tourism studies with journalism studies. Tourism scholars Steve Pan and Cathy Hsu focus on the ways in which travel journalism contributes to the formation of destination images – those mental maps that tourists have of a place. Their analysis focuses on the coverage of the top five destinations for Mainland Chinese and how they were covered in Chinese travel magazines. They argue that travel journalists rarely challenge dominant frames about a foreign destination, and are complicit in furthering stereotypes.

Part II of this book is concerned with the producers of travel journalism, whether they are professionals employed in mainstream media organizations, or amateur and student travel journalists producing content online. In Chapter 5, Bryan Pirolli takes a close look at evolving practices in travel blogs about Paris. The chapter is particularly interested in evaluating the relationship between such blogs and traditional journalistic standards and practices. Going beyond merely the producers, he also explores the interpretations and expectations of those who read these blogs. Chapter 6 also focuses on the way in which the digital environment is affecting travel journalism. Andrew Duffy examines its implications for travel journalism students, who might be pre-conditioned by accessing travel journalism online. He argues that it is important to challenge students to go beyond mere re-telling of experiences to engage meaningfully with host nations. Concentrating on India, Usha Raman and Divya Choudary explore in Chapter 7 the motivations of amateur travel journalists; they look at both travel blogs and newspapers, thus straddling old and new media platforms. A key argument in their analysis is that, while in newspaper travel journalism content tends to be homogenized, the open-endedness of travel blogs and their potential to build communities outside the commercialism of the industry means they offer new and unique ways to practice travel journalism. Chapter 8 explores travel journalism in the context of another populous Asian country that is sending increasing numbers of tourists around the world. Jiannu Bao looks at the evolution of travel journalism in China, and argues that we can discern three stages of this development, which are reflective of the broader evolution of the Chinese media system. She maps in great detail the evolution from a propaganda function to one focused mainly on personal expression and alternative voices.

In Part III, contributors explore aspects of the content of travel journalism more closely. Folker Hanusch's comparative analysis of travel stories in newspapers from Australia, Britain, Canada and New Zealand, presented in Chapter 9, asks whether the coverage of travel is actually a significant departure from foreign news reporting, in terms of the countries covered. His findings demonstrate that the opportunity for travel journalism to present a more balanced view of the world is generally not taken up, with significant similarities between travel journalism and foreign news. In Chapter 10, Ben Cocking directs his analysis to British travel journalism about safari holidays in Africa. He is concerned especially with the representational strategies that journalists use to report on these experiences, and finds that the commercial environment in which they are told means that stories are geared towards audiences' cultural expectations. Thus, travel journalism is unable to break free from clichéd and at times outdated views of the world. Chapter 11, by Christine Buzinde, Eunice Yoo and C. Bjørn Peterson, continues the regional analysis, though this time with a focus on the Middle East. Employing textual analysis, their contribution highlights the visual and verbal discourses about the region as portrayed through the popular television travel program *No Reservations*. They demonstrate that travel journalism can confront audiences with socio-political issues in a way that engages them by strategically involving and empowering local perspectives.

The final part of this book focuses on the politics of travel journalism through a variety of prisms. In Chapter 12, Wiebke Schoon explores how the concept of cosmopolitanization can be used to elucidate travel journalism. This mainly theoretical chapter offers a very useful and concrete framework that operationalizes cosmopolitanism in a way that can be adapted for future content analyses of travel journalism. This is an important departure, as inquiries of cosmopolitanism in the media have tended to concentrate on disaster and crisis reporting, rather than softer types of journalism. Lyn McGaurr extends her work on the role of travel journalism in communicating environmental problems in Chapter 13. Also taking a cosmopolitanist approach, she provides an empirical analysis of travel journalists' views of their reporting on the environment in the Australian island state of Tasmania. Her analysis demonstrates that while travel journalism can show at times cosmopolitan concern for destinations, it is also still tied very strongly to the market logic of the global tourism industry, making such representations the exceptions rather than the norm. Chapter 14 focuses on a phenomenon that has been explored in tourism studies for only a relatively short amount of time. Brian Creech explores the ways in which travel journalism can mediate dark tourism: travel to sites of disasters or death. He analyzes the reporting of the Tuol Sleng prison in Cambodia, which is now a museum. He argues that travel journalism, while mostly remaining superficial, can connect with audiences to evoke empathy and humanism in different ways from news journalism.

(Travel) journalism in turbulent times

The debate continues if travel journalism can present a unique and significant perspective on how we understand the world. While some of the authors in this book find travel journalism reiterating or intensifying the problems of traditional news journalism, others demonstrate under what conditions this type of journalism can challenge our knowledge or augment our perspectives derived from international news. If anything, the chapters detail that a more balanced understanding of the world's diversity does not come easily and needs self-reflective and innovative professional strategies. Credibility, transparency and active engagement in close encounters – those are the journalistic ingredients in various chapters that trigger empowering discourse, interrupt problematic representations and offer more than what traditional news journalism can provide.

Yet there are no quick and easy solutions to better journalism. This book then urges travel journalists to push the boundaries of their field, to leave the pack on easy trips organized by tourism public relations agencies. At a time when the media industry is in upheaval and the profession of journalism fears extinction, this task seems daunting. But these troubled times also provide travel journalists with ample opportunities by offering more technological capabilities, media platforms, and narrative options than ever. However, renewal does not just mean looking ahead but may also entail remembering well-established practices of travel journalism. The foreign and (sometimes) travel correspondent Robert D. Kaplan even sees traditional strategies of travel journalists and writers as a model for journalism in general:

> Journalism desperately needs a return to *terrain*, to the kind of first-hand, solitary discovery of local knowledge best associated with old-fashioned travel writing. Travel writing is more important than ever as a means to reveal the vivid reality of places that get lost in the elevator music of 24-hour media reports. (2006, p. 49, emphasis in original)

Travel journalism is a field with unique circumstances but also an area that echoes the trials and tribulations of the media industry overall. This book hopes to show that studying travel journalism does not just help explain a journalistic niche field but can also clarify professional assumptions and practices of other types of journalism. The findings from the fringes can have consequences for the study of all journalism and the media in general. It is about rethinking and revitalizing journalism at a time of crisis.

References

Austin, Elizabeth (1999) 'All Expenses Paid: Exploring the Ethical Swamp of Travel Writing', *The Washington Monthly*, 31.7/8, pp. 8–11.

Beerli, Asunción and Martín, Josefa D. (2004) 'Factors Influencing Destination Image', *Annals of Tourism Research*, 31.3, pp. 657–81.

Bruns, Axel (2005) *Gatewatching: Collaborative Online News Production*, Peter Lang: New York.

Burns, Peter M. and Novelli, Marina (eds) (2009) *Tourism and Politics: Global Frameworks and Local Realities*, Oxford: Elsevier.

Cohen, Erik (1979) 'A Phenomenology of Tourist Experiences', *Sociology*, 13.2, pp. 179–201.

Flew, Terry (2012) *The Creative Industries: Culture and Policy*, London: Sage.

Florida, Richard (2012) *The Rise of the Creative Class: Revisited*, New York: Basic Books.

Fürsich, Elfriede (2002) 'Packaging Culture: The Potential and Limitations of Travel Programs on Global Television', *Communication Quarterly*, 50.2, pp. 204–26.

Fürsich, Elfriede (2003) 'Between Credibility and Commodification: Nonfiction Entertainment as a Global Media Genre', *International Journal of Cultural Studies*, 6.2, pp. 131–53.

Fürsich, Elfriede and Kavoori, Anandam P. (2001) 'Mapping a Critical Framework for the Study of Travel Journalism', *International Journal of Cultural Studies*, 4.2, pp. 149–71.

Graburn, Nelson, H.H. and Barthel-Bouchier, Diane (2001) 'Relocating the Tourist', *International Sociology*, 16.2, pp. 147–58.

Hanitzsch, Thomas (2007) 'Deconstructing Journalism Culture: Toward a Universal Theory', *Communication Theory*, 17.4, pp. 367–85.

Hanusch, Folker (2012a) 'Broadening the Focus: The Case for Lifestyle Journalism as a Field of Scholarly Inquiry', *Journalism Practice*, 6.1, pp. 2–11.

Hanusch, Folker (2012b) 'A Profile of Australian Travel Journalists' Professional Views and Ethical Standards', *Journalism*, 13.5, pp. 668–86.

Hanusch, Folker (2012c) 'Travel Journalists' Attitudes toward Public Relations: Findings from a Representative Survey', *Public Relations Review*, 38.1, pp. 69–75.

Hartley, John (1996) *Popular Reality: Journalism, Modernity, Popular Culture*, London: Arnold.

Hill-James, Candeeda R. (2006) *Citizen Tourist: Newspaper Travel Journalism's Responsibility to Its Audience*. Unpublished M.A. thesis, Brisbane: Queensland University of Technology.

Jaworski, Adam and Pritchard, Annette (eds) (2005) *Discourse, Communication and Tourism*, Clevedon: Channel View.

Kaplan, Robert D. (2006). 'Cultivating Loneliness: The Importance of Slipping Away from the Pack to Encounter, and Understand, the World Firsthand', *Columbia Journalism Review*, January/February, pp. 48–51.

McNair, Brian (1998) *The Sociology of Journalism*, London: Arnold.

Mercille, Julien (2005) 'Media Effects on Image: The Case of Tibet', *Annals of Tourism Research*, 32.4, pp. 1039–55.

Nielsen, Christian (2001) *Tourism and the Media: Tourist Decision-Making, Information and Communication*, Elsternwick: Hospitality Press.

Plog, Stanley (2001) 'Why Destination Areas Rise and Fall in Popularity: An Update of a Cornell Quarterly Classic', *Cornell Hotel and Restaurant Administration Quarterly*, 42.3, pp. 13–24.

Reeves, Geoffrey (1993) *Communications and the 'Third World'*. London: Routledge.

Ross, Andrew (2009). *Nice Work If You Can Get It: Life and Labor in Precarious Times*, New York: New York University Press.

Santos, Carla Almeida (2004) 'Framing Portugal: Representational Dynamics', *Annals of Tourism Research*, 31.1, pp. 122–38.

Schudson, Michael (2003) *The Sociology of News*, New York: W.W. Norton.

Scott, Julie and Selwyn, Tom (eds) (2010) *Thinking through Tourism*, Oxford: Berg.

Smith, Valene L. (ed.) (1989) *Hosts and Guests: The Anthropology of Tourism*, Philadelphia: University of Pennsylvania Press.

Smith, Valene L. and Brent, Maryann (eds) (2001) *Hosts and Guests Revisited: Tourism Issues of the 21st Century*, New York: Cognizant Communication Corp.

Swick, Thomas (1997) 'On the Road without a Pulitzer', *The American Scholar*, 66.3, pp. 423–29.

Thurlow, Crispin and Jaworski, Adam (2010) *Tourism Discourse: Language and Global Mobility*. New York, NY: Palgrave Macmillan.

Towner, John (1995) 'What Is Tourism's History?' *Tourism Management*, 16.5, pp. 339–43.

TripAdvisor (2013) *TripBarometer by TripAdvisor*. Available at: http://www. tripadvisortripbarometer.com/download/Global%20Reports/TripBarometer% 20by%20TripAdvisor%20-%20Global%20Report%20-%20USA.pdf.

Urry, John (1990) *The Tourist Gaze: Leisure and Travel in Contemporary Societies*, London: Sage.

Urry, John (1995) *Consuming Places*, London: Routledge.

Urry, John and Larsen, Jonas (2011) *The Tourist Gaze 3.0*. London: Sage.

Waters, Somerset R. (1997) *Travel Industry World Yearbook: The big picture—1996–1997, Volume 40*. New York: Child &Waters.

World Tourism Organization (UNWTO) (2012a) *International Tourism Hits one Billion*. (Press release, 12 December). Available at: http://media.unwto.org/en/ press-release/2012-12-12/international-tourism-hits-one-billion.

World Tourism Organization (UNWTO) (2012b) *UNWTO Welcomes the World's One-Billionth Tourist*. (Press release, 13 December). Available at: http://media.unwto. org/press-release/2012-12-13/unwto-welcomes-world-s-one-billionth-tourist.

World Tourism Organization (UNWTO) (2013) *UNWTO Tourism Highlights 2013*. Available at: http://dtxtq4w60xqpw.cloudfront.net/sites/all/files/pdf/ unwto_highlights13_en_hr.pdf.

World Tourism Organization UNTWO (2014) *International Tourism Exceeds Expectations with Arrivals Up by 52 Million in 2013*. (Press Release, 20 January). Available at: http://media.unwto.org/press-release/2014-01-20/ international-tourism-exceeds-expectations-arrivals-52-million-2013.

World Travel & Tourism Council (2013) *Economic Impact 2013 World* [Online]. Available: http://www.wttc.org/site_media/uploads/downloads/world2013_1. pdf.

Zelizer, Barbie (2004) *Taking Journalism Seriously: News and the Academy*, Thousand Oaks: Sage.

Zelizer, Barbie (2011) 'Journalism in the Service of Communication', *Journal of Communication*, 61.1, pp. 1–21.

Part I
Mapping the Terrain: Strategies for Studying Travel Journalism

2
People on the Move: Travel Journalism, Globalization and Mobility

Elfriede Fürsich and Anandam P. Kavoori

Introduction

Travel journalism is an important site for studying the ideological dimensions of transcultural encounters, mobility and the ongoing dynamics of globalization. This chapter provides a theoretical and programmatic framework for the investigation of contemporary travel journalism in the broader context of international communication and globalization studies. We see travel journalism as an institutional site where meaning is created and where a collective version of the 'Other/ We' is negotiated, contested and constantly redefined.

Based on scholarship in sociology, anthropology, geography and tourism studies, along with cultural and communication studies, we present four interrelated theoretical perspectives for the analysis of travel journalism, which are structured by issues of periodization, power, experience and mobility. In particular, we revisit and update our article 'Mapping a critical framework for the study of travel journalism' (Fürsich and Kavoori, 2001), which more than ten years ago initiated a stream of research in this area and, ultimately, contributed considerably to this book being published.

First, the section on periodization discusses how the cultural ideas of modernity and postmodernity have shaped our understanding of travel and tourism. We argue that the rise and decline of nationalism as a central political category challenges the typical 'us versus them' dichotomies of travel journalism. Second, the section on power and identity highlights the ways the seemingly trivial field of travel journalism ties into issues of cultural imperialism. In particular, we explain how travel journalism exemplifies the problematic practice of identity formation *ex negativo* – in contrast to an Other. Third, in a section on experience and authenticity we discuss dynamics of tourist and host interactions related to typologies of tourists/tourism and the quest for authenticity in tourist practices. Fourth, a new section on mobility explains how scholarship on travel journalism can productively utilize recent contributions from the so-called mobility paradigm.

Under each of these headings, we first map the various conceptual issues that might frame our understanding of tourism and travel journalism before then identifying research questions that studies on travel journalism might address. We suggest examining issues of both encoding (including studies of journalistic work routines or media practices, and the textual analysis of travel journalism) and decoding, or reception. Moreover, we present research that has already tackled some of these issues.

Issues of periodization

Modernity and postmodernity as historic periods and cultural experiences help explain attitudes toward tourism. Closely related are advances in nationalism and postnationalism, which find their theoretical equivalent in discussions surrounding cosmopolitanism.

Modernity

It is not easy to identify the complex cultural frameworks that have made tourism the enormous sociological fact that it is. For some, the answer lies in the emergence of tourism as an offshoot of modernity. Two issues peculiar to modernity can be seen as underlying the growth of tourism: first, the creation of work and leisure as separate spheres of social activity and the division of the social environment into discrete units of experience; and second, the impact of technology on everyday life.

Urry (1990, pp. 2–3) argues that tourism's location within hours of leisure time (rather than structured work time) makes it a truly modern enterprise: 'It is one manifestation of how work and leisure are organized as separate and regulated spheres of social practice in "modern" societies'. Modernity has been also characterized by transformed cultural dynamics, as rural communities – with their strong association with traditional sources of cultural coherence (e.g., religion and patriarchy) – were replaced by large urban centers characterized by a search for alternate sources of cultural coherence. At the heart of this search was the desire for authenticity. As MacCannell (1976, p. 3) explains, 'for moderns, reality and authenticity are thought to be elsewhere: in other historical periods and other cultures, in purer, simpler lifestyles'.

MacCannell sees the artificial preservation, reconstruction or the muse-umization of traditional socio-cultural forms (for example, the 'primi-tive' or the 'exotic') as the ultimate sign that modernity has succeeded. Tourism becomes the 'spirit' of modernity.

In addition, any modernist approach to analyzing travel cannot overlook how modern technology, travel and tourism are intrinsically linked. For example, the tourist 'gaze' is often mediated through a cam-era (first photography and later video) as Urry explains. In its extreme, 'travel is a strategy for the accumulation of photographs' (1990, p.139). This photogenic logic structures tourist locations (vistas) and social experience. Professional travel photography and television travel shows can be understood as an extension of this motivation.

Each of these issues resonates with scholarly interest in travel journal-ism. At the level of production and textualization, it would be useful to ask questions about the exact conjuncture of the development of travel journalism with the rise of modernity. The history of the newspaper is tied to industrialization and the notion of work itself. Concomitantly, it would be interesting to explore changing notions of leisure and recrea-tion within early newspapers. Since colonial narratives usually endorse travel as an educational activity, one should analyze how the modernist idea of leisure as an activity in and for itself develops in relation to its colonial origins.

Postmodernism

While the literature on postmodernism connects with tourism in a number of ways, one issue has direct relevance for scholars of tourism and travel journalism: the hybridity of cultural forms in postmodernity. For example, Leong argues:

> Tourist culture in effect is a showcase of postmodernism: a concoc-tion of something 'native,' something borrowed, something old and something new. ... In such a melange, authenticity is somewhat eclipsed by estrangement: dance and other rituals, organized for tourist consumption, become performances rather than integral parts of the social life of participants. (1989, p. 371)

Urry also calls the 'tourist gaze' a symbol of postmodernity. Like Leong, he sees typical aspects of postmodernism (pastiche, kitsch, the hyper-real, copies without original, dissolution of boundaries) represented in tourism. For him, 'being a tourist' is the quintessential mode of post-modern experience.

[T]he era of mass communication has transformed the tourist gaze and many of the features of postmodernism have already been partly prefigured in existing tourist practices. ... [P]eople are much of the time 'tourists' whether they like it or not. The tourist gaze is intrinsically part of contemporary experience, of postmodernism. (1990, p. 82)

Viewing tourism through the prism of postmodernity, provokes us to ask whether there is the possibility of a postmodern travel journalism that blurs the boundaries between 'real' and 'staged' authenticities. Textual analyses can examine the different subjectivities portrayed in the range of tourist periodicals, newsletters, television shows and Internet sites. At the level of reception, researchers can ask: What is the role of mass tourist periodicals in a postmodern age that encourages the development of differentiated and hybrid identity forms? What is the range of specific cultural (and subcultural) mediations of these already fragmented texts?

Nationalism

Modernity is inextricably linked to nationalism and the formation of the nation-state (Giddens, 1990). To become 'modern' is often seen as a goal by emergent nation-states. Within this context, tourism is understood as fulfilling the goals of economic development of the nation-state and thus 'modernizing' it. However, tourism has more than just an economic function. In many cases, tourism's function is symbolic, working to create a map of the nation for both internal and external consumption. In the United Kingdom, for example, the tourist industry 'sells' the monarchy and its imperial legacy as prime tourist attractions along with the 'pristine' English countryside with its village inns and cricket greens. Such tours construct an image of England and the English that essentializes a certain narrowly prescribed notion of nationhood based on a particular identity (white, male and upper class). As Stuart Hall (1989) points out, such a discursive rendering of culture is implicitly ideological and has been less reflective of England's multiculturalism than it is of the cultural ideology of political conservatism.

Tourism works in similar ways in other parts of the world. Leong (1989) shows how states – in his case Singapore – manufacture traditions for tourism, by choosing to emphasize certain areas of the past in the process of nation building. National tourism becomes a site for the construction of unified national symbols, even if this involves the 'cleansing' of cultural practices that are thought of as unattractive to tourists. Instead of presenting existing varied lifestyles, Singapore creates a new 'exotic' that – combined with Western amenities and facilities – becomes the travel package 'Instant Asia': 'Since most visitors stay no more than three days in Singapore, marketers of cultural meanings try to show and sell those messages quickly in a condensed form' (p. 366).

Yet, as the tourist industry (public or private) generates national traditions and constantly refers to past nostalgic sites, the modernization process may be thwarted in counterproductive ways. Chang and Holt discuss the dilemma facing Taiwan:

> Forcing a culture into the straitjacket of a defined 'past' serves to fix cultural Others in a 'timeless present,' confining them to a place that cannot change, or cannot change as easily as the representer's culture. (1991, p. 116)

This dilemma is not limited to Asian or European countries. Rowe (1993), for example, points to the Australian travel industry's juxtaposition of nostalgia for nature ('the bush') and the past against modern urban life ('the city') – a construction that does not work smoothly. Comparing it to British travel images, he concludes, 'whereas Britain is romanticizing its past as civilization, Australia is marketing the timelessness of nature. This is ironic for a nation allegedly seeking to modernize itself' (p. 264). This irony is intensified, as Rowe points out, by the fact that tourism always brings about major changes in infrastructure and therefore threatens the very existence of undisturbed nature.

A recent urgency lies in the fact that nation-states as political entities and nationalism as a unifying ideology have come under pressure in an era of political, economic and cultural globalization. While the degree of this dissolution and the rise of postnationalism are still debated (Breen and O'Neill, 2010), a reaction to this development has been a renewed interest in cosmopolitanism as a philosophical idea, political principle and sociological practice (Appiah, 2006; Beck, 2006; Delanty, 2009). Schoon (Chapter 12) and McGaurr (Chapter 13) demonstrate in this book how concepts of cosmopolitanism can be productively employed as a framework for the analysis of travel content or as concept for theorizing the ideological underpinnings of this type of journalism.

The entire process of nationalism and its discursive construction provides an important site for studying travel journalism. At the level of production and textualization, the questions are numerous: Where does travel journalism fit into agendas of nationalist construction and postna-tionalist dissolutions both historically and in contemporary times? How do travel services (agencies, newspaper sections, television shows, public relations) tie into discourses of nationalism or cosmopolitanism? Whose vision of nation is being served by each of these institutions? How does travel journalism in developing nations reconcile the paradoxes between modernization, authenticity and change? At the level of recep-tion analysis one can investigate how the vision of the nation and its discursive destinations in travel journalism direct actual tourist practices, for example at national monuments (such as the Vietnam Memorial or India Gate)? Over time, how has travel journalism produced or rejected various dimensions of cosmopolitanism?

Power and identity

Studies of international communication and culture have emphasized issues of power through two main approaches – cultural imperialism on the one hand and ideology and identity formation on the other. Both these approaches are useful for understanding tourism and travel jour-nalism in its cross-cultural context.

Cultural imperialism

Hall (1994) suggested 20 years ago that tourism research in marketing and economics, with its 'value-free' approach to science, has failed to consider the political context of tourism. He contends that

> tourism is a central element of some of the critical economic and political issues of the contemporary era: the internationalization of capital; industrial and regional restructuring; urban redevelopment; and the growth of the service economy. (Hall, 1994, p. 197)

Hall questions the naive approach that sees tourism as a 'force of peace' and a non-polluting industry, as many policy makers try to promote it. Instead, as theorists working from a dependency framework have argued, tourism performs economic functions for the

> rich tourist-generating super powers ... it reinforces rather than undermines the structures of dependency and underdevelopment upon which the world system exists. Active encouragement of tour-ism by developing countries has led to an enmeshment in a global system over which they have little control. (Britton, 1982, p. 331)

More recently, issues of sustainability and environmental protection have become major causes for concern in tourism development (Mowforth and Munt, 2008). The tourism industry cannot escape echoing existing economic and political inequalities.

Nevertheless, while a tremendous transfer of money, labor and goods takes place within this largest business sector in the world, its impact goes well beyond monetary effects. As MacCannell (1976) writes, 'not only our old favorite, "the profit motive," operates unambiguously in the development of [tourist] attractions. Some attractions are developed and maintained at great expense, though there are no economic returns' (p. 162). MacCannell's work suggests that tourism should not be reduced solely to a form of economic and political dependency but should also be considered with regard to its cultural and ideological aspects. For some, however, tourism then becomes a form of cultural domination constructed by the West – a new form of cultural imperialism (e.g., Nash, 1989). The question becomes to what extent travel journalism is implicated in cementing unequal geopolitical relationships. Pratt (1992), who has examined 19th-century travel writing, offers a model for the discursive analysis of the representational practices and politics of this genre. We can extend Pratt's approach to contemporary travel journalism: At the level of textual analysis, one can ask what the dominant frames are, and what modes of representation are at work, in Western travel journalism. Who benefits from such frames and who does not? What are the changes that take place in travel representations with changing geopolitical interests of nation-states? Can travel journalism be seen as an ongoing expression of cultural transgression or does it simply reflect the current inequalities between the global North and South? Research on travel-media representations of destinations in crisis areas such as the Middle East, offers examples of such transgressions, and the possibility of representations beyond those common in news journalism (see Buzinde, Yoo and Peterson, Chapter 11; Cocking, 2009).

At the level of textualization and reception, however, any theorizing of tourism as one-sided dominating ideological transfer overlooks the complexity of this cultural practice and discourse. Integrating Foucault's (1980) and Spivak's (see Parry, 1987) notions of multiplicity, complicity and multiple articulations, Pratt develops a more interdependent idea for the practice of travel writing. She employs two concepts ('contact zone' and 'transculturation') useful for our purposes. 'Contact zones' refer to the space of colonial encounters ... a contact perspective emphasizes how subjects are constituted in and by their relations to each other. It treats the relations between colonizers and colonized, or travelers and 'travelees' not in terms of separateness or apartheid, but in terms of co-presence, interaction, interlocking understandings and practices, often within radically asymmetrical relations of power. (Pratt, 1992, p. 7)

With the term 'transculturation', Pratt highlights the general process of cultural mediation of this relationship and, specifically, the cultural texts produced by such encounters. These texts are not only the Eurocentric texts of the Western (or First World) travel writers, but equally, the texts of the Third World (or historically, of the colonized) that appropriate or even resist the originals.

Transferring these questions to contemporary journalism and media reception, we need to study how the images of a Western Other are incorporated by the 'others' themselves. Moreover, what is the range of travel journalism texts produced in the Global South? Are there oppositional strategies for constructing travel destinations that do not adhere to the discursive strategies of Western texts? Ultimately, we need to recognize that tourism is not a one-way stream from the 'First' to the 'Third' World. A growing number of people (not just elites) travel to the Global North as tourists, students and workers. Traveling abroad has become a major activity for new middle classes in emerging economies. The question how these new tourism constituencies and their associated journalism construct travel experience is at the center of Raman and Choudary's study on Indian travel bloggers (Chapter 7 in this book).

Ideology and identity formation

Issues of ideology or identity formation and maintenance have been at the forefront of cultural studies, and they have relevance for the study of tourism and travel journalism. Generally speaking, ideology formation from the perspective of cultural studies is seen as a symbolic process. Like metaphors for language, ideologies are indispensable for the construction of meaning. They provide a framework for action and for understanding complex and incomprehensible realities – in the case of tourism the contact with the Other. Like a road map, ideologies provide at once a model of and for reality (Geertz, 1973). Therefore, the analysis of the ideological dimension of tourism involves all aspects of symbolic processes. From such a perspective, travel journalism can be considered a key site of ideological formation, one that functions through contrasting practices. As Urry (1990, p. 1), for example, writes, the gaze in any historical period is constructed in relationship to its opposite, to non-tourist forms of social experience and consciousness'.

Contrasting travel journalists' accounts with underlying assumptions of the opposite allows for an understanding of what is taken for granted. Debates in the field of cultural anthropology illuminate this point. Chang and Holt (1991, p. 103) explain how closely connected the experiences of ethnographers and tourists are: 'Both ethnographers and tourist engage in cultural inscriptions (not descriptions) that lead them to fixate the host culture in a certain way'.

These ideas generate another set of questions for evaluating travel journalism. At the level of encoding, we can ask what ideological work do travel journalists perform? It appears that travel journalists, even more than 'average' tourists, are trying to fix the Other. Their professional purpose is to come up with a narrative, a well-told story about other cultures, the past or distant places – in short, to package culture. This places travel journalists in a liminal position; they operate at the border between the foreign and the familiar. This position makes their discourse especially charged. Studies of travel journalism need to consider the complexity of how this narrative is created. Moreover, at a time when globalizing trends in business and media have resulted in a 'global identity crisis' (Morley and Robins, 1995, p. 10), this position 'on the border' puts travel journalists in the critical position of cultural translators. The process of defining identity takes place across many sites. As Rowe (1993, p. 261) explains:

> National, racial-ethnic as well as metaphysical mythologies are constructed and selectively mobilized in the act of consumer persuasion. These images are not, therefore, mere inventions or devices, nor can their ramifications be limited to the tourist cash nexus. They form part of the ensemble of cultural, economic, social and political relations in play in the constitution of society.

Travel journalism, much like tourism, is a multifaceted cultural practice, which involves many stakeholders: government, private travel industry, travel writers/journalists, tourists, public relations agencies and advertising firms. Research needs to address how these agencies – all with different and even contradictory interests – collectively shape an image of the Other. Numerous issues can be addressed: What kinds of 'identity work' do travel industries, as part of the cultural industry, perform for audiences? To what extent do the texts of travel journalism produce identity across social categories such as class, race and gender? Given that travel journalism is about selling destinations and cultures, are 'positive' identities replacing 'negative' images of the Other? Several studies have illuminated the complex and problematic 'othering' practices unique to travel journalism such as exoticization (Fürsich, 2002, 2010).

Experience and authenticity

Research in sociology, anthropology and tourism studies has examined the specific experiences and dynamics of tourist and host interactions, and has often formulated typologies of tourists or tourism and tourist experiences. Smith (1989), for example, divides tourism into five varieties (ethnic, cultural, historical, environmental and recreational) while Cohen (1973) provided an earlier typology of four tourist roles – the organized mass tourist, the individual mass tourist, the explorer and the drifter. The first two are categories of institutionalized tourist roles that prefigure greater social distance between tourists and locals, and the latter two are non-institutionalized roles and reflect meaningful cross-cultural interaction. Cohen (1979) later classifies five modes of tourist experiences. Based on a phenomenological classification of tourist motivations, he distinguishes between recreational, diversionary, experimental, experiential and existential modes.

The issue of authenticity

In his now-classic analysis, MacCannell (1976, p. 1), drawing on the work of Durkheim and Goffman, examines tourism as a form of 'staged authenticity' and experiences in 'touristic spaces'. Tourism, MacCannell argues, is a modern form of the essentially religious quest for authenticity, but tourists are often robbed of that experience. The hosts create 'tourist spaces', in which spurious attractions are presented as if they were 'real'. In other words, purveyors of culture 'stage authenticity' for tourist consumption. The hosts strategically create tourist spaces when they allow tourists to encroach on the 'back' regions of the culture. Such staging of roles is especially evident with mass tourism, where tourists become dehumanized objects to be tolerated for economic gain (Pi-Sunyer, 1977) and subject to excessive familiarity or hostility, because of a failure to understand social roles within different cultures (Smith, 1989).

Travelers themselves tend to distance themselves from 'tourism'. MacCannell calls this the dilemma of anti-tourism versus pro-tourism, and suggests that:

Tourists are not criticized ... for leaving home to see sights. They are reproached for being satisfied with superficial experiences of other people and places. ... [T]ouristic shame is not based on being a tourist but on not being tourist enough, on a failure to see everything the way it 'ought' to be seen. (1976, p. 10)

Instead of successfully challenging the shortcomings of mass tourism, anti-tourists present a reified version of 'modern' tourism as unachievable model. One group of anti-tourists, which has not given up on travelling, but contests the standards of traditional mass tourism, is introduced by Corrigan (1997), who uses the term *untourists*, after an Australian travel network established in the early 1990s. These untourists are analogous to 'eco tourists, green tourists, cultural tourists, alternative tourists, educational tourists' (according to an activist quoted in Corrigan, 1997, p. 144). These tourists often travel with an environmental mission, as a reaction to the problem of mass tourism. They still are searching for authenticity (beyond the impossibilities of the postmodern age) but it is an elite construct of authenticity, high on cultural capital and distinctively class defining. Quite a few untourists deliberately reject luxury at their destination after spending a large amount of money to reach it in the first place (e.g., wilderness trek in the Amazon region) (1997, p. 145).

Tomaselli adds a semiotic perspective to the problematic of tourism to indigenous people in Africa (by Western tourists, filmmakers or anthropologists). He uses the term cultural tourism to distinguish this form of tourism as the 'commodification of difference' (2001, p. 176). In several case studies, Tomaselli shows the symbolic work of Western tourists, tour guides and native populations in Africa while creating the tourist experience. Often locals actively 'reinvent their authenticity' (1999, p. 188) to create a tourist experience. All parties together seem to develop agreed upon representations of the tribes that are accepted by the tourists as authentic experience but often have nothing to do with the actual, more westernized lives of these tribes. Certain moments in Tomaselli's case studies even hint at more postmodern forms of tourism, similar to what Feifer (1985) elucidates. *Post-tourists*, according to Feifer, actively play with the inauthenticity of the modern tourist experience. They 'know that there is *no* authentic tourist experience, that there are merely a series of games or text that can be played' (quoted in Urry, 1990, p. 11). Similarly, MacCannell (2008) points to the need of active co-production between hosts and guests for a moment of staged authenticity to be considered successful.

Corrigan (1997, pp. 145–6) evaluates the different newer types of tourism from the often-overlooked perspective of class and anticipates:

> It seems likely that post-tourists and untourists will be the elite tourist categories of the future, the former representing the postmodern strand of culture ... and the latter growing out of the environmental strand. Ironic players with appearance and earnest seekers after 'authentic' reality: the two poles of philosophical approaches to the world hardly ever seem to change.

Rojek reaches a step further and argues that the 'quest for authenticity is a declining force in tourist motivation' (1997, p. 71). Instead, he sees tourists actively create meaningful experiences in processes of symbolic 'indexing and dragging'. In a media-saturated world, being a traveler and mobility become a standard state of mind – no longer a distinction from everyday life. As the media make everything seem familiar, difference needs to be culturally and artificially created (1997, p. 71).

The research questions linking travel journalism to issues of experience and authenticity could consider how travel journalism structures tourist experience. What concepts of ideal tour and preferred traveler are created? How does the journalists' engagement with the hosts impact their understanding of tourism? In which forms of tourism do most travel journalists engage (based on various typologies)? At the level of reception, one can study the specific dynamics between travel journalism and tourist institutions. This would entail examining the range of 'cultural brokers' that mediate tourism sites. Joseph (1994), for example, found that local hosts in an Indian tourist town knew all the major travel writers and actively cultivated them, thereby actively configuring the tourist experience.

The issue of authenticity is one that lends itself well to the study of both the historical and contemporary texts of travel journalism. What are the constructs of authenticity that travel journalism has predominantly used in the past and now? At the level of reception, one can ask: What are the modes of authenticity to which travelers most subscribe? What textual strategies work best to further issues of authenticity and leisure? Is there a tension between the textualization of authenticity and the actual experience?

The centrality of the concept of authenticity for any analysis of travel journalism is evident in this book, in which several authors tackle the issue (for example, Duffy, Chapter 6; Pirolli, Chapter 5).

Tourism, travel and mobilities

Another promising approach to understanding travel journalism is provided by recent scholarship on mobility. Issues of mobility are obviously at the center of theorizing tourism and travel. Advancements in modern transportation technology initiated modern mass tourism, and they continue to guide tourism experience. Prato and Trivero (1985, p. 40), for example, show how the cultural history of the car, train and plane has structured the ideological aspects of travel and tourism, up to a point when motion and transport 'ceases to function as a metaphor of progress or at least of "modern" life, and becomes instead the primary activity of existence'.

More recently, the 'mobility turn' has changed the perspectives of various disciplines. According to Urry (2007), mobility research 'examines how social relations necessitate the intermittent and intersecting movements of people, objects, information and images across distance' (p. 54). Especially scholars in sociology and cultural geography have taken concepts of mobility as a starting point for a new research perspective. All these approaches have in common an acceptance that mobility, and the motivation (or need) to be mobile, are intrinsically linked to structures of contemporary life.

Mobility itself is a complex phenomenon. Researchers have named various definitions and dimensions (e.g., Adey, 2010; Cresswell, 2006; Urry, 2007); however, in general, most forms of mobility can be connected to three types. First, *physical* mobility describes people on the move (migrants, tourists, business travelers, commuters) but also the movement of goods, resources and capital in space and time. Second, *social* mobility designates the more traditional understanding in sociology, which relates to the ability to change social position and status. Third, *virtual* mobility involves the movement of ideas, images and information in symbolic spheres.

Past decades have brought intensive engagement with the ideas of the global, the transcultural and the hybrid. By intersecting these concepts with insights of mobility studies we can further expand our repertoire of theories and methodologies for understanding global media flows in general and travel media in particular.

The advantage of mobility as a concept is that it moves research on globalization beyond symbolic, political and regulatory aspects to a more pragmatic level of the lived experience of people in mobility-driven and technology-saturated environments. Moreover, mobility studies more seriously engage with the materiality and symbolic endowments of space.

Tourism scholarship on creating, branding or experiencing destinations is an important foundation for understanding the creation of place (Sheller and Urry, 2004). It is no coincidence that research that tries to capture mobile practices focuses on what Urry (2007) called 'transfer cites', such as airports, stations, motels or waiting rooms – all significant locations of tourism.

Mobility research also directs our attention to the producers of media content. Communication scholars have focused on audiences across the world, but have often overlooked the dimension of mobility among media workers themselves. Journalists traverse borders – as travel journalists, foreign correspondents and war correspondents, but also whenever their work crosses class and ethnic divides. All along, they work on cross-cultural constructions of 'us' and 'them'.

The mobility turn, moreover, offers new approaches to research practice. Scholars in sociology and cultural geography integrate innovative research methodologies (e.g., Urry, 2007, pp. 39–42; Adey, 2010, p. xv) that account for the increasing mobilization of society. Of interest for the study of the reception of (traditional and emergent) travel media are new research technologies that attempt to capture the mobility of research subjects by using positioning technology such as GPS or mobile phones for space–time diaries and other forms of dynamic research designs. In addition, virtual methodology such as link mapping and geographies of web architecture can help in analyzing the complexities of travel content on the Internet.

Questions of identity and belonging remain a focus of many types of mobility analyses. The most significant problem seems to be the nonexistence of stable structures. A world on the move – people, goods and information – requires us to examine the underlying cultural assumptions of these long-term processes. Here a more historic perspective on travel journalism content promises constructive insights. Just as colonial travel writing has been read as evidence for ideologies of Othering and dominance, so can popular travel journalism of the past decades be investigated for its ideological assumption on mobility, progress or cultural diversity.

In addition, scholars have begun to challenge all-too-enthusiastic celebrations of mobility as a *zeitgeist* and the 'fetishism of movement and mobility' (Urry, 2007, p. 197) that may forget its drawbacks. This work confronts the social, political and ethical aspects of the need to be mobile and the individual and social costs of mobility (for example, Bergmann and Sager, 2008). Closely related to these problems is the question of the environmental price we pay to support a mobile world. Sustainability of mobility has become an urgent issue that an increasing number of mobility scholars now take seriously (Holden, 2007) and that the UN World Travel Organization lists as a major concern (World Tourism Organization, 2013).

Of special interest are also the consequences of immobility. For example, both Urry (2007) and Kaufmann (2002) interrogate the new obstacles that mobile societies create. Kaufmann introduces the helpful concept of 'motility' as 'the way in which an individual appropriates what is possible in the domain of mobility and puts this potential to use for her or his ability' (2002, p. 37). His studies illustrate how social and spatial aspects of mobility are interdependent. Weighing the social and individual aspects that make mobility possible, he operationalizes mobility as a compromise between someone's access to mobility, the potential competence and the actual appropriation by making use of mobility. Urry (2007) develops the concept of 'network capital' to outline the impediments that need to be overcome for access to a mobile society. For him, owning network capital is 'the capacity to engender and sustain social relations with those people who are not necessarily proximate and which generates emotional, financial and practical benefit' (p. 197). Aspects of this capital range from access to the right travel documents and mobile technologies to a network of people that can be visited for personal or professional gain.

Travel journalism studies infused with concepts of mobility research can draw attention to several questions. On the encoding site, research can ask how travel journalists engage in the field. Participant observation can help examine the intricacies of professional practice of travel journalists as workers on the move. Mobile tracking technology can delineate journalists' movement on and off the beaten track. Results can be interesting for other border-crossing journalists as well. The content produced by travel journalists for various media offers an abundance of historic and contemporary material to track ideologies of mobility and progress. Gabriele (2006), for example, combined dimensions of historic travel journalism, mobility and gender to a fascinating analysis of the relationship of women's lives and national identity. Recent travel content can be interrogated for its negotiation of issues of motility, sustainability and surveillance at a time when demands of mobility have come under scrutiny. McGaurr's studies (2010; Chapter 13) on the impact of environmental problems and sustainability on travel journalism are exemplary.

On the side of reception, it will be interesting to study how travel information changes when it is 'mobilized' with mobile technology. The fact that an increasing amount of travel information is accessed virtually, and used and produced by professional and amateur writers while on the move, requires analyses of the impact of mobility on content and reception experiences. Moreover, scholars can also examine how consuming travel journalism and using other travel content contributes to 'network capital' and 'motility' and stipulates or hinders mobility.

Beyond travel as cultural metaphor

It is striking that cultural metaphors of travel have been popular to explain the unique situation of living in modernity and in postmodernity. From such mobile figures as Benjamin's 'flaneur' (Hanssen, 2006) as the quintessential modern person, to Bauman's invoking of 'the stroller, the vagabond, the tourist and the player' (Bauman, 1995, p. 91) as postmodern counterparts, nomadism and traveling have been used to make sense of the ideologies and structures of feelings that guide contemporary lives. By interrogating actual travel practices and travel journalism from a cultural perspective, we help to ground some overly celebratory meanings of travel and tourism.

More than ten years after the first version of this essay, researchers interested in travel media can draw on a recent boom in sociological (Apostolopoulos et al., 2001) and anthropological (Burns, 2004) research on tourism. At the same time, the questioning of mobilities challenges the purpose of travel and tourism and its affiliated journalistic practices. The tourism market is more differentiated than ever from the hyper-packaged all-inclusive vacation to the luxurious (or Spartan) approaches of the 'ecotourist' to the sustainable mobility of the 'untourist'. Similarly, travel media are more complex, globalized and fragmented. The platforms for travel journalism and content have increased tremendously since 2001 through globalization and digitalization. And, as several chapters in this book describe, the influence and uniqueness of professional writers has been challenged by amateur bloggers and online reviewers. Analyzing travel journalism has become more complicated but also more multifaceted. The profession and genre was one of the first that saw its authority undermined by digital technology, as travel websites were amongst the first sites that were economically viable. Scholars should observe whether travel journalists might also be among the first ones who find new legitimation and even business models in the crisis surrounding traditional journalistic approaches.

Our aims for the chapter have remained primarily programmatic – to map a theoretical framework for studying travel journalism. By calling attention to and theorizing this genre, we hope to encourage journalism and media scholars, as well as others, to begin to understand the global mobilities that shape all our lives whether or not we are currently on the move.

References

Adey, Peter (2010) *Mobility*, London: Routledge.

Apostolopoulos, Yiorgos, Leivadi, Stella and Yiannakis, Andrew (eds) (2001) *The Sociology of Tourism: Theoretical and Empirical Investigations*, Oxon, UK: Routledge.

Appiah, Kwame Anthony (2006) *Cosmopolitanism: Ethics in a World of Strangers*, New York: W.W. Norton.

Bauman, Zygmunt (1995) *Life in Fragments*, Oxford: Blackwell.

Beck, Ulrich (2006) *The Cosmopolitan Vision*, Cambridge: Polity Press.

Bergmann, Sigurd and Sager, Tore (eds) (2008) *The Ethics of Mobilities: Rethinking Place, Exclusion and Environment*, Surrey: Ashgate.

Breen, Keith and O'Neill, Shane (2010) *After the Nation? Critical Reflections on Nationalism and Postnationalism*, Hampshire, UK: Palgrave Macmillan.

Britton, Stephen G. (1982) 'The Political Economy of Tourism in the Third World', *Annals of Tourism Research*, 9.3, pp. 331–58.

Burns, Georgette Leah (2004), 'Anthropology and Tourism: Past Contributions and Future Theoretical Challenges', *Anthropological Forum*, 14.1, pp. 5–22.

Chang, Hui-Ching and Holt, Richard (1991) 'Tourism as Consciousness of Struggle', *Critical Studies in Mass Communication*, 8.1, pp. 102–18.

Cocking, Ben. (2009) 'Travel Journalism: Europe Imagining the Middle East', *Journalism Studies*, 10.1, pp. 54–68.

Cohen, Erik (1973) 'Nomads from Affluence: Notes on the Phenomenon of Drifter-Tourism', *International Journal of Comparative Sociology*, 14.1-2, pp. 89–102.

Cohen, Erik (1979) 'A Phenomenology of Tourist Experiences', *Sociology*, 13.2, pp. 179–201.

Corrigan, Peter (1997) *The Sociology of Consumption: An Introduction*, London: Sage.

Cresswell, Tim (2006) *On the Move: The Politics of Mobility in the Modern West*, London: Routledge.

Delanty, Gerard (2009) *The Cosmopolitan Imagination: The Renewal of Critical Social Theory*, Cambridge: Cambridge University Press.

Feifer, Maxine (1985). *Going Places: The Way of the Tourist from Imperial Rome to the Present Day*. London: Macmillan.

Foucault, Michel (1980) *Power/Knowledge: Selected Interviews and Other Writings, 1972–77 (edited by Colin Gordon)*, New York: Pantheon Books.

Fürsich, Elfriede (2002) 'Packaging Culture: The Potential and Limitations of Travel Journalism on Global Television', *Communication Quarterly*, 50.2, pp. 203–25.

Fürsich, Elfriede (2010) 'Media and the Representation of Others', *International Social Science Journal*, 61/issue 199, pp. 113–30.

Fürsich, Elfriede and Kavoori, Anandam P. (2001) 'Mapping a Critical Framework for the Study of Travel Journalism', *International Journal of Cultural Studies*, 4.2, pp. 149–71 [doi: 10.1177/136787790100400202].

Gabriele, Sandra (2006) 'Gendered Mobility, the Nation and the Woman's Page: Exploring the Mobile Practices of the Canadian Lady Journalist, 1888–1895', *Journalism*, 7.2, pp. 174–96.

Geertz, Clifford (1973) *The Interpretation of Cultures*, New York: Basic Books.

Giddens, Anthony (1990) *The Consequences of Modernity*, Stanford, CA: Stanford University Press.

Hall, Colin Michael (1994) *Tourism and Politics: Policy, Power and Place*, Chichester, UK: John Wiley.

Hall, Stuart (1989) 'New Ethnicities', in Kobena Mercer (ed.), *Black Film, British Cinema*, London: ICA Documents 7, pp. 27–31.

Hanssen, Beatrice (ed.) (2006) *Walter Benjamin and the Arcades Project*, London; New York: Continuum.

Holden, Erling (2007) *Everyday and Leisure-Time Travel in the EU*, Aldershot: Ashgate.

Joseph, Christina (1994) *Touts, Tourists and Tirtha: The Articulation of Sacred Space at a Hindu Pilgrimage,* unpublished Ph.D. Dissertation, University of Rochester, New York.

Kaufmann, Vincent (2002) *Re-Thinking Mobility: Contemporary Sociology*, Aldershot: Ashgate.

Leong, Wai-Teng (1989) 'Culture and the State: Manufacturing Traditions for Tourism', *Critical Studies in Mass Communications*, 6.4, pp. 355–75.

MacCannell, Dean (1976) *The Tourist: A New Theory of the Leisure Class*, New York: Schocken.

MacCannell, Dean (2008) 'Why it Never Really Was about *Authenticity'*, *Sociology*, 45, pp. 334–7.

McGaurr, Lyn (2010) 'Travel Journalism and Environmental Conflict: A Cosmopolitan Perspective', *Journalism Studies*, 11.1, pp. 50–67.

Morley, David and Robins, Kevin (1995) *Spaces of Identity: Global Media, Electronic Landscapes and Cultural Boundaries,* London: Routledge.

Mowforth, Martin and Munt, Ian (2008) *Tourism and Sustainability: Development, Globalisation and New Tourism in the Third World*, 3rd edn, Oxon: Routledge.

Nash, Dennison (1989) Tourism as a Form of Imperialism, in Valene L. Smith (ed.), *Hosts and Guests: The Anthropology of Tourism*, 2nd edn., Philadelphia: University of Pennsylvania Press, pp. 37–52.

Parry, Benita (1987) 'Problems in Current Theories of Colonial Discourse', *Oxford Literary Review*, 9(1/2), pp. 27–58.

Pi-Sunyer, Oriol (1977) Through Native Eyes: Tourists and Tourism in a Catalan Maritime Community, in Valene L. Smith (ed.), *Hosts and Guests: The Anthropology of Tourism*, first edn., Philadelphia: University of Pennsylvania Press, pp. 149–56.

Prato, Paolo and Trivero, Gianluca (1985) 'The Spectacle of Travel (translated by Iain Chambers)', *Australian Journal of Cultural Studies*, 3.2, pp. 25–42.

Pratt, Mary Louise (1992) *Imperial Eyes: Travel Writing and Transculturation*, London: Routledge.

Rojek, Chris (1997) 'Indexing, Dragging and the Social Construction of Tourist Sights', in Chris Rojek and John Urry (eds), *Touring Cultures: Transformations of Travel and Theory*, London: Routledge, pp. 52–74.

Rowe, David (1993) 'Leisure, Tourism and "Australianness"', *Media, Culture & Society*, 15.2, pp. 253–69.

Smith, Valene L. (1989) Introduction, in Valene L. Smith (ed.) *Hosts and Guests: The Anthropology of Tourism*, 2nd edn., Philadelphia: University of Pennsylvania Press, pp. 1–17.

Sheller, Mimi and Urry, John (eds) (2004) *Tourism Mobilities: Places to Play, Places in Play*, London: Routledge.

Tomaselli, Keyan G. (1999) 'Psychospiritual Ecoscience: The Ju/'hoansi and Cultural Tourism', *Visual Anthropology*, 12.2–3, pp. 185–95.

Tomaselli, Keyan G. (2001) 'The Semiotics of Anthropological Authenticity: The Film Apparatus and Cultural Accommodation', *Visual Anthropology*, 14.2, pp. 173–83.

Urry, John (1990) *The Tourist Gaze: Leisure and Travel in Contemporary Societies*, London: Sage.

Urry, John (2007) *Mobilities,* Cambridge: Polity.

World Tourism Organization (UNWTO) (2013) *World Tourism Organization: Sustainable Development of Tourism Programme.* Available at: www.sdt.unwto.org.

3
Armchair Tourism: The Travel Series as a Hybrid Genre

Maja Sonne Damkjær and Anne Marit Waade

Introduction: travel journalism and lifestyle media

Travel journalism is increasing in magazines, newspapers and online media and on TV, representing a significant change in news and entertainment media over the past few decades. Lifestyle journalism and non-fiction entertainment are growing because they attract audiences, advertisements and product placement while at the same time they are easy to plan and cheap to produce for editors and broadcasters (Christensen, 2010; From, 2007). Travel and tourism constitute an essential part of contemporary Western popular lifestyle and indicate social and cultural status and preferences (Dunn, 2005a; Urry and Larsen, 2011). This has resulted in an increase in travel journalism, on the Internet, in newspapers and magazines and on television, where – along with other lifestyle programs on topics including food, interior design, fashion, house, garden, health and social change – travel programs are used for channel branding and securing market shares (Johnson, 2012). Within this media landscape, travel series comprise an important and yet unexplored field of study. The term 'armchair tourism' indicates that travel and tourism have become popular media entertainment and parts of a widespread mediated lifestyle consumer culture (Waade, 2006).

In this chapter we want to categorize the travel series as a hybrid genre. Our main questions are which genre elements constitute the travel series, and what are the cultural dimensions of the travel series as a hybrid journalistic genre. The theoretical section of the chapter includes a discussion of key international contributions to the field. The empirical section encompasses a record of all travel series shown on six Danish television channels in the period 1988–2005.[1] These series are used to illustrate and discuss the theoretical perspectives. Based on our study, we present a tentative typology of 10 different travel series subgenres. We argue that the travel series is a rapidly transforming and complex hybrid genre linked to such diverse fields as (lifestyle) journalism, television entertainment, travelogues and tourism. Finally, we discuss some current challenges for travel series, and how they influence the genre's characteristics.

The travel series as media entertainment and consumer information

Travel series offer anything from travel documentary to consumer guidance, exotic entertainment and vivid visual material for 'spatial phantasmagoria' (Jansson, 2002, p. 435). In some contexts, 'travel series' solely refers to the travel documentary genre, while host-based lifestyle TV entertainment is labeled as 'holiday programs', and the term 'travel show' is used to describe consumer and tourism guidance programs. The travel series in its basic form is positioned between three central text genres: journalistic documentary, factual television entertainment and consumer information,[2] as illustrated in Figure 3.1.

Each of these three genres entails a specific communicative intentionality (Scannell, 1994, p. 41) and mode of representation (cf. Nichols, 2001). The primary intent of consumer information is to persuade and influence the viewer in relation to consumer behavior, while the

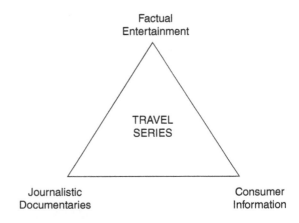

Figure 3.1   Travel series

journalistic documentary appeals to the recipient as a democratic citizen; the primary purpose of television entertainment is to capture and retain the viewers' interest by offering aesthetic pleasure, humor, excitement and audience involvement. In relation to these three categories, certain subtypes of travel series have developed over the course of the last 25 years. Our analytic triangle can be used to place and characterize new hybrid program types within the genre and delineate different TV channels' travel series profiles.

In the Danish context, the heyday of the travel series was in the 1990s and the early 2000s. Since then, travel series have mainly been produced for niche channels in order to distinguish the channels in an increasingly competitive television market. In a global context, travel series are produced for an international market mainly based on the logics and values of the tourism and media industry; the general tendency in this context is that in these programs commercial and entertainment values replace journalistic ideals (Fürsich, 2002b; Waade, 2009). There is a fundamental difference between addressing audiences as citizens and addressing them as consumers (Hanitzsch, 2007). Nevertheless, one can argue that even factual entertainment and soft news enable the viewers to relate to their social and cultural environment as cultural citizens through enjoyment, absorption and consumption.

Research perspectives on travel journalism

Lifestyle media and global television entertainment have received extensive research attention over the past decade (Bell and Hollows, 2005; Bonner, 2003; Hill, 2007; Palmer, 2008). But when it comes to studies of travel media as entertaining and sponsored journalism, the contributions are far more limited. Hanusch (2010) presents a comprehensive review of research contributions on travel journalism in an international context. He highlights four dimensions in existing research and identifies several shortcomings in the field. His research represents a classical journalistic approach in which ideals of free press, critical journalism, news criteria and source criticism are paramount. His four dimensions identify relevant discussions in relation to travel journalism in general and the travel series in particular, but they do not capture the hybrid genre nature of the travel series or the importance of media-specific features. Likewise, travel journalism's significance as entertainment remains unexplored. While this news-journalism perspective on travel journalism is important, we call for a broader perspective to understand travel journalism as a cultural phenomenon.

Nordic research on travel journalism and the travel series

In the Nordic context, research on travel series and travel journalism primarily focuses on how foreign cultures and countries are represented. For example, Orgeret's (2002) analysis of the Norwegian travel series, *Gutta på tur* (Boys on a trip) discusses specific national discourses that are negotiated in the travel series as a genre. A related approach investigates national discourses and constructions of 'the other' in news media (Eide and Simonsen, 2004). Similarly, Kabel (2005) has conducted an extensive analysis of news, factual programs and travel series' coverage of the Third World on Danish television. According to Kabel's study, viewers of travel series prefer dramaturgic conflicts rather than news journalism conflicts. Kabel highlights humor, identification and drama as essential ways to engage viewers in relation to foreign cultures:

> While news coverage ... was about the West's confrontation with the Arab world and Islam, the factual documentary series showed a wide and vivid view on lives and world conditions; they brought the world closer to the Danish viewers. (2005, p. 66)

It may seem odd that Kabel's study compares travel programs with news coverage from war zones. Of course, the contextualization of locations makes a difference, for example, as part of the 'developing world' or as an 'exciting tourism destination'. However, the analysis illustrates the variety of intentions and functions that factual television can fulfill.

It may seem paradoxical that Hanusch demands more critical angles in travel journalism, while Kabel calls for news and documentary series that are less conflict oriented – more pleasurable and entertaining. One of the reasons for this contradiction is that Kabel and Hanusch investigate the travel journalism field from very different angles. What is lacking in both of them is a more general questioning of travel journalism as a cultural field. Travelogues, travel journalism and international news reports have some features in common, but they are based on fundamentally different journalistic and genre traditions, motivations and premises. For example, the travelogue is a literary genre based on a narrative structure, travel journalism is feature journalism covering travel as lifestyle and consumer culture, and finally, international news reports are based on journalism structured by news criteria (for example, a conflict's significance and identification) and cover a wide range of issues from political, cultural and social conditions. Furthermore, editorially the content is located in different media sections, for example in weekly magazines of newspapers, foreign news sections in news media or non-fiction television entertainment. In our view, travel journalism is primarily to be seen as lifestyle journalism with certain significance and premises. Lifestyle journalism is characterized by its specific lifestyle-related topics (for example interior design, gardening, cooking, fashion, health, travel, education), its lack of journalistic conflicts (though it sometimes includes dramaturgic conflicts) and its friendly tone and atmosphere (Carlsen and Frandsen, 2005).

The lack of critical perspectives on foreign cultures and consumer culture within travel journalism can be seen as a problem. The only critical element often relates to consumer guidance on where to find the cheapest or best place to go, and where *not* to go. A deeper critical journalistic engagement is generally missing. An interesting case in this context is DR1's (the oldest public broadcaster in Denmark) cancellation of Karsten Kjær's costly travel series *Rejsen til Paradis* (The Journey to Paradise, 2005) due to its uncritical depiction of, among other things, the totalitarian state of Burma and famine-hit Ethiopia as tourist paradises.

Feel-good journalism: pleasurable images, practical guidance and celebrities

To investigate the entertaining aspects of travel series, it is helpful to do so in the context of research on the broader field of lifestyle journalism. From (2007; From and Waade, 2007) examined lifestyle journalism in Danish newspapers, and the relationship between consumer journalism in a historical perspective and today's more lifestyle-oriented service journalism. According to From, the basic appeal of lifestyle journalism is its usefulness and the enjoyment it provides. *Usefulness* is about receiving practical information and guidance as a consumer, while *enjoyment* includes the series' positive, aesthetic and sensory-based descriptions stimulating the viewer's emotions or imagination (Jantzen and Vetner, 2008; Waade, 2010). From and Kristensen (2011) describe lifestyle journalism as a kind of cultural journalism; 'cultural' in this context includes sport, art, consumer information, media entertainment and tourism. According to the authors, the characteristic aim of culture and lifestyle journalism in current Danish newspapers is to amuse, entertain and update, and to create identities and communities among citizens (From and Kristensen, 2011, pp. 193ff).

Travel series include significant entertainment and media-specific characteristics. In addition to the fascinating and pleasurable images of exotic places, peoples and cultures, as well as attractive holiday experiences and practical travel guidance, the host is an important aspect of the series (Fürsich, 2002a). As in many different factual television formats, the host is the one who ensures sociability, credibility, humor and a positive atmosphere. The host addresses viewers directly, introduces the locals, participants, guests and viewers to each other, and guides everybody through the program. This feel-good television has been mixed with other formats and entertainment elements such as quiz, reality and makeover, which is also the case in the travel series (Dunn, 2005a, 2005b). The host is in many cases a celebrity, thus obliging audience interest in gossip and insight into celebrities' personal and professional lives – another feature of entertainment (Rojek, 2001).

To sum up, we argue that the travel series is a form of feel-good journalism that should be considered in the context of other lifestyle journalism in which readers/viewers are addressed as customers, hedonists and consumers, rather than citizens (From, 2007). The travel series is an audio-visual subgenre of travel journalism that engages the viewer through drama, humor and identification. Further categorization asks for considering media-specific characteristics, including the role of the host, types of journeys, destinations and elements from other TV genres. The travel series is a hybrid genre that mixes elements from other genres and discursive practices, and it is the mixing ratio that characterizes travel series and other factual television entertainment as media phenomena (Hill, 2007).

The travel series in a genre perspective

In our study of travel programs we draw on the functional perspective of genre that has emerged in media studies, especially in the study of television programs (Andersen, 1994; Hill, 2007). The core of this conception is that genre is established through systems of orientations, expectations and conventions that circulate between producers, texts and audiences (Neale, 1980, p. 19). This perspective marks a shift from a text-centered approach to genre analysis towards an understanding of genre in terms of function and purpose (Andersen, 1994, p. 17). The outcomes have been multiple studies that combine genre analysis with production and reception studies (Bruun, 2010; Frandsen and Bruun, 2005). In line with this audience awareness, we understand travel journalism as a genre linked not only to journalistic practices and conventions, but also to expectations of entertainment and consumer guidance. Documentary theorist Nichols' (2001) study of documentary film style has been an important inspiration for our framework. He offers a conceptual model with six different 'modes of representation' aiming to distinguish particular traits and conventions of various documentary film styles. The modes are not mutually exclusive; on the contrary, in practice the modalities often overlap significantly between specific documentary films. Nichols presents a flexible concept of 'dominant mode', which means that the characteristics of a given mode can function as dominant in a given film without dictating or determining every aspect of it. He draws the contours of a chronological development of the modes but emphasizes that, in practice, documentary films often revert to themes, styles and devices from previous modes (Nichols, 2001, p. 100). While Nichols' genre perspective is focused on the text, media scholar Hill (2007) turns towards the audience in her genre analysis of factual television; but both authors suggest a dynamic view of genre and genre work. In accordance with Nichols' description of the documentary film genre, Hill (2007, p. 212) points out that hybridity is a distinctive feature of factuality, and television documentaries are characterized therefore by a balance and blending of different genre elements. We have incorporated Nichols' and Hills' dynamic notion of genre in our analysis of travel series.

Methodological design

We based our analysis on qualitative content analysis. This type of analysis is primarily inductive and emphasizes an integrated view of texts and their contexts. Qualitative content analysis explores meanings, patterns and themes in the text, and it can be used to generate theory (Zhang and Wildemuth, 2009; Hsieh & Shannon, 2005). The process of qualitative content analysis we used involves a set of systematic procedures that resembles the steps in traditional quantitative content analysis (e.g. Berelson, 1952): prepare data, define the unit of analysis, develop and test coding scheme, code the text, asses coding consistency, draw conclusions and report methods and findings (Zhang and Wildemuth, 2009).

As unit of analysis, we chose to limit our data set to travel programs broadcast in Denmark, as many travel series are produced with a local language community in mind, which is reflected in the choice of well-known hosts and participants. An assessment of these aspects requires a thorough knowledge of the cultural community from which these media texts emerge and in which they are embedded. Our data set consisted of travel series broadcast in prime time during 1988–2005 on the public service channel DR1 and its sister channel DR2, on the partly commercial public service channel TV2 and its commercial sister channels TV2 Zulu and TV2 Charlie and, lastly, on the commercial channel TV3. The year

1988 was chosen as the starting point for our data collection because this was when DR, Denmark's independent, license-financed, national broadcasting corporation lost its monopoly on national television, and TV2 was launched. TV2 soon became popular, and it now holds a larger share than DR1 (the former television channel DR). Despite the launch of new TV-channels during the 1990s, the two public service channels, DR1 and TV2, still dominate the Danish TV market and attract more than half of the daily audiences.

We only included travel series aired in prime time, meaning programs starting between 7–10 pm, since throughout the 1990s and early 2000s this time of day was packed with TV programs inspired by lifestyle journalism. Programs in this time slot are typically designed with entertainment in mind, and this is particularly noticeable on the two public service channels in the early evening hours (Christensen, 2008, p. 23). Our delimitation meant that travel series aired in the morning or afternoon, typically slots reserved for factual information or education, were not included. Early in our research process we discovered that seriality proved to be an essential characteristic of television travel programs, which would separate 'travel series' from 'foreign affairs programs'. Single television programs and classic journalistic features about foreign countries or international affairs, which typically consist of single-themed features, were removed from our dataset, and we included only travel series consisting of at least two episodes. Moreover, we discarded series that did not include travel, tour guiding or a touristic gaze (Urry and Larsen, 2011) as a constituent concept element. Following these selection criteria, we arrived at a dataset of 146 travel series spanning a period of 17 years.

As a base for our analysis we initially coded and counted selected content in the travel series. We developed a coding scheme built on categories generated from our review of previous studies and theories in the field, as explained in more detail below. To validate our coding scheme, we developed a coding manual, and revised coding rules, until we achieved consistency.[3] After an initial coding of a sample of our data, we checked for coding correspondence, and discussed and resolved any ambiguities concerning definitions of categories, coding rules or categorization of specific cases. We checked coding repeatedly during the process and after having coded the entire data set. On this basis, we started the iterative process of analyzing, comparing and grouping different content elements; this was followed by an interpretation of patterns and unifying characteristics, until we reached agreement on a tentative typology of travel series.

In order to gain a basic understanding of key concepts and production characteristics in our data set, we screened and coded for a number of carefully selected parameters. Given that the host is an important factor for the entertaining and phatic qualities of lifestyle TV formats, we coded for hosting (whether or not the show hosted), the gender of the host, and the number of hosts per episode and show. In addition, we considered and recorded whether the host could be characterized as a 'well-known media personality' in relation to the channel's audience, as this is important for the series' entertainment aspects. Further, we listed travel type, by distinguishing between shows that focus on a round trip or a particular destination, or in which the travel form and destination is eclipsed by social interaction. Finally, in order to clarify the origin of the travel series, we coded for production country. Based on these coding criteria, and the main theme of the different series, we evaluated whether the series were organized according to the dominant principles of journalistic documentary, factual television entertainment, consumer information or mixed forms.

Analysis of the travel series

Our coding procedure made it possible to establish patterns by comparing hosting design, travel focus and origin of production with the series' theme and dominant text genre. This resulted in ten types of travel series covering the variations of the dataset. The result of the coding is illustrated in Figure 3.2, which shows the overall categorization of the travel series broadcast in prime time on DR1, DR2, TV2, TV2 Zulu, TV2 Charlie and TV3 during the years 1988–2005. Our typology follows Nichols' concept of 'dominant mode': certain characteristics dominate the individual travel series without dictating every aspect of it. Therefore, there can be variations in practice that do not meet all of the characteristics of a certain type. However, the majority of the travel series in our dataset met all the characteristics of the type to which they had been ascribed.

As a prelude to a more detailed description of the ten travel series types that emerged from our analysis, we illustrate how they relate to the three text genres of journalistic documentary, factual television entertainment and consumer information, on the basis of Figure 3.2.

In the lower left corner of the triangle, where the journalistic principles dominate, we find the types 'Travelogue' and 'Popular Science'. Here the destination or the round trip is in focus, and the role of the narrator or host is primarily to enlighten and engage viewers in the

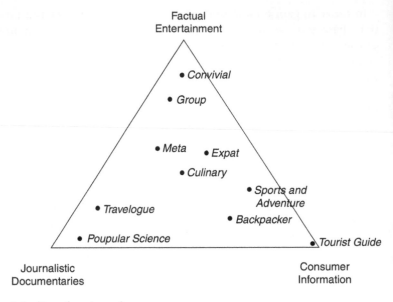

Figure 3.2   Travel series sub-genres

foreign culture and place. At the top of the triangle, the principles of factual television entertainment dominate. This is where we find 'Group' and 'Convivial', offering soft-scripted, plotted formats that prioritize social banter between participants or hosts, while they discover exotic destinations. To the right of the triangle, the principles of consumer information and advertising dominate. Here we find the types 'Tourist Guide', 'Sports and Adventure' and 'Backpacker'. They all have explicit consumer information as an essential part of their format, for example evaluations of lodging or local attractions. In the center of the triangle the types 'Culinary', 'Expat' and 'Meta' are located, as they are not dominated by either of the vertices, but balance a mixture of the three main genres. Before a discussion of our findings, we will present each of the ten travel series types in more detail.

Ten types of travel series

The most common travel series type in the material is 'Travelogue', which is molded by journalistic documentary, orchestrated as a round-trip with a specific project in mind (for example an anthropological or a biological mission), and usually host driven. Typically, the host is a well-known male, and, as illustrated in Table 3.1, this combination clearly takes precedence over all other variations of hosting types. The host acts as a tour guide, gives lectures on location and engages with nature, the local culture and population. In this way, 'Travelogue' shows resemble the documentary mode Nichols has characterized as participatory (2001). A classic example of a 'Travelogue' series is BBC's *Himalaya with Michael Palin*, aired on the public service channel DR1 in 2004. The Travelogue constitutes a majority of travel series broadcast on DR1, 67 per cent, as shown in Table 3.2, which gives the percentage of different travel series types aired on each of the six Danish channels during the years 1988–2005. The table also points to the 'Travelogue's' status as a classic: with the exception of TV2 Zulu, which was launched in 2000 as a commercial venture aiming at entertaining the 15–30 age group, the 'Travelogue' comprises an important type on all the Danish channels, but extensively so on DR1.

Closely linked to the text genre of journalistic documentaries, we also find another type: 'Popular Science'. This type typically focuses on the destination, for example in terms of exploring a specific geographic area's nature, natural phenomena, wildlife or history, as seen in, for example, *The Silk Road*, aired on TV2 in 1989. 'Popular Science' is often produced in accordance with Nichols' definition of the expository mode (2001, p. 105). This mode of documentary emphasizes verbal commentary and argumentative logic and is usually driven not by a visible host but by a voice-over supported by evidentiary editing, meaning that images are used to illustrate or substantiate what is said. In the beginning of the period under study, the type 'Popular Science' mainly appeared on the public service channels, DR1 and the partly commercial TV2, but it has since then become a dominant travel series type on the 'highbrow' public service channel DR2, launched in 1996.

A completely different type of travel series is produced as factual television entertainment, where social interaction in combination with visual pleasure is favored and usually dominates the series' focus on travel form or destinations. Linked to this text genre we find the type we termed 'Convivial', which is an umbrella term for a wide variety of formats that include elements of game show, reality and makeover television formats with travel as a staple element. 'Convivial' programs are particularly common on the commercial entertainment channel TV3, launched in 1990, on the youth-focused TV2 Zulu, and to some extent on TV2. As a central characteristic, celebrities or unknown participants are placed in an arranged or staged situation abroad – typically in some form of social experiment (for example, two families swapping

Table 3.1 Results of coding and analysis: travel series aired on DR1, DR2, TV2, TV2 Zulu, TV2 Charlie and TV3, 1988–2005

DOMINANT TEXT GENRE	TRAVEL SERIES TYPES	NUMBER OF SERIES	No host	Host – Male Celebrity	Host – Male Unknown	Host – Female Celebrity	Host – Female Unknown	Two or more hosts per episode* Celebrity	Two or more hosts per episode* Unknown	Changing hosts in the series* Celebrity	Changing hosts in the series* Unknown	PRODUCTION COUNTRY**	FOCUS Round trip	FOCUS Destination	FOCUS Social interaction
Journalistic Documentary	*Travelogue*	46		36	7	2	1					UK: 11 DK: 33 US: 1 SE: 1	37	9	
	Popular Science	21	15	3	1				m/f: 1		m/f: 1	UK: 3 Co: 1 AU: 1 US: 1 FR: 1 DK: 14	5	15	1
Factual Entertainment	*Convivial*	19	1	1				m/f: 10 m/m: 1 f/f/f: 1	m/f: 2	m/f: 4		DK: 18 UK: 1			19
	Group	2	1	1								DK: 2	1		1

Consumer Orientation								Gender*		Production country**			
Tourist Guide	2	1						m/f: 1		DK: 1 / DE: 1			2
Backpacker	13					m/f: 2			m/f: 11	US: 2 / UK: 11	11		2
Sports and Adventure	12	4	2	1	1		f/f: 1	m/m: 1 / m/f: 1	m/f: 1	DK: 11 / Co: 1			12
Mixture													
Expat	3	2		1						DK: 3			3
Meta	2	1				m/m/m: 1				DK: 1 / Co: 1	2		
Culinary	26	12	4	8				m/f: 1	m/f: 1	DK: 11 / UK: 14 / AU: 1	2	24	
Total	*146*	*20*	*57*	*14*	*4*	*10*	*17*	*6*	*4*	*14*	*146*	*58*	*55*	*33*

*Gender: m: male, f: female.
**Production country: Co: Co-production, DK: Denmark, SE: Sweden, UK: United Kingdom, AU: Australia, DE: Germany, FR: France, US: United States.

Table 3.2 Distribution of the travel series types aired on DR1, DR2, TV2, TV2 Zulu, TV2 Charlie and TV3, 1988–2005

TYPES	CHANNELS										
	DR1		DR2		TV2		TV2 Zulu		TV2 Charlie		TV3
	N	%	N	%	N	%	N	%	N	%	N	%
Travelogue	23	67	10	17	9	30			3	25	1	20
Popular Science	5	15	14	24	1	4			1	9		
Convivial	1	3	2	4	7	23	5	72	1	8	3	60
Group					1	3			1	8		
Tourist Guide	1	3					1	14				
Backpacker			13	23								
Sports and Adventure			6	10	5	17					1	20
Expat					1	3	1	14	1	8		
Meta	2	6										
Culinary	2	6	13	22	6	20			5	42		
N (Total – 146)	34		58		30		7		12		5	

holidays). This is the concept of *Rejsefeber* (Travel Fever), aired in 2005 on the advertising-funded TV 2 Charlie, a channel that was launched one year earlier with a mature audience in mind. Other formats belonging to this type include game elements and center around celebrities carrying out different tasks while they discover an exotic location, which is the case in *Eventyrerne* (The Adventurers), aired on TV2 Zulu in 2002. The 'Convivial' frequently has more than one host per episode, and it is often produced locally, to a national audience that knows the hosts, the celebrities and the cultural codes that come into play on the show.

A less common type of travel series, which can also be categorized as factual television entertainment, we call 'Group Travel'. As the name suggests, the viewer follows a larger group, typically a family, on a long trip. This is the case in *Familie på farten* (Family on the go) aired in 2005 on TV2 Charlie. In 'Group Travel' the surroundings are largely the backdrop of a collective grand tour, and the focus of attention is on the social interaction and the development of the groups in connection to their adventures in a foreign country and culture.

The destinations play a much more prominent role in another less widespread type of travel series we call 'Tourist Guide', which falls under the text genre of Consumer Information. A central element of this type of series is the sequence of picturesque sceneries and sensuous images accompanied by background music. Here the host is typically replaced by a voice-over extolling the pleasure, value and enjoyment of specific destinations or types of vacation. An example is *Drømmerejser* (Dream Vacations) aired on DR1 in 1989. The program presents highlights from luxury vacation settings such as expensive hotels, cruises or resorts on remote islands. Although only a few examples of this type of travel series are present in our empirical material, it is a type of travel series common on international TV channels like Travel Channel.

The text genre of consumer information also includes the type 'Backpacker', which appears solely on the 'highbrow' public service channel DR2 and accounts for around a quarter of the channel's travel series. 'Backpacker' shows follow independent travelers on low-cost journeys to foreign places. Similar to the 'Travelogue' it has a participating and outreaching host as its narrative engine, but in contrast, the backpacker travel series is not as dependent on the host's persona, but rather focuses on travel rituals and culture. Moreover, 'Backpacker' programs have different hosts of both sexes who, rather than being well-known media personalities prior to their appearance in the series, are often experienced backpackers. The 'Backpacker' type usually displays a relatively clear consumer journalism angle that complements the host's engagement in the local culture and search for the ambience of the destination 'off the beaten track' with an evaluation of low budget experiences, dining and accommodation. It might be seen as a paradox that 'Backpacker' shows are closely related to consumer information while backpacker culture celebrates authenticity and avoidance of consumerism and the mainstream tourism industry; but backpacker culture has become an industry itself with specific consumer values, favorite destinations and forms of traveling (Sørensen, 1999). Production companies such as Pilot Guides, which also publish guidebooks and host large travel sites online, have specialized in this international travel series format; accordingly, there were no specific Danish productions of this type in our material.

Another consumer-oriented type of travel series we termed 'Sports and Adventure', which was aired on TV3, TV2 and to a lesser extent on DR2, and has a varied constellation of hosts, as shown in Figure 3.2. 'Sports and Adventure' revolves around the cult of sports and outdoor activities, and the destination is presented and evaluated on the basis of its suitability as a backdrop for specific activities. This is the case in *På ski igen* (Going Skiing Again), which has a skiing instructor as its host, and which aired on TV2 in 2000.

In addition to the seven types of travel series described so far, we found another three in the cross field between the text genres journalistic documentary, factual television entertainment and consumer information. The most common is the type 'Culinary', which resembles the 'Travelogue', but here the host is a chef – often a well-known male gourmet. It revolves around food, meals and local cuisine, typically within a specific geographic area, as the name of a classic example, Floyd on OZ, aired on DR2 in 1997, indicates. With its focus on food culture, this type of travel series is widespread, especially on TV2 Charlie, the channel for a mature audience, where it accounts for over 40 per cent of the travel series aired in the investigated period (Table 3.2).

In the cross field we also find the 'Expat' as a type that portrays individuals or families who have emigrated and the destination they have settled in. An example is *Når ude er hjemme* (Living Abroad) aired on TV2 Charlie in 2005. The series presents personal stories about everyday life in different foreign destinations presented as authentic tourism information through a native perspective. The last travel series type we found, 'Meta', is especially self-reflexive or presents a parody of the genre. It can be semi-fictional, for example when the host takes on another character; or it can be 'autofictional', meaning that well-known media personalities play the roles as themselves (Jacobsen, 2008). The latter is the case in the Scandinavian co-production *I sandhedens tjeneste* (In Search of the Truth), aired on DR1 in 1991, which was an experimental travel series about Europe. In many ways, 'Meta' series resemble Nichols' description of a documentary in the reflexive documentary mode, which questions the authenticity of documentary in general (2001, pp. 125–137).

Overall, in this section we have argued that we need to categorize travel journalism, and especially the travel series, on the basis of other parameters than the traditional principles of news journalism. We included parameters that account for the central features of factual television entertainment, such as hosting and travel mode. Based on our analysis we argued that the travel series is a varied journalistic hybrid genre developed in the field between journalistic documentary, factual television entertainment and consumer information. Since we confined our sample to travel series aired on Danish TV channels, we acknowledge that there might be further travel series types present in other broadcasting systems. Nevertheless, it is probable that the travel series types we have identified here are also found in other countries. With the exception of the type 'Group', at least one series of each of the types analyzed here was produced outside of Denmark.

Discussion: travel as cross-media and sponsored entertaining journalism

The media landscape has changed radically over the past decade. A general trend in the Nordic context is that the travel series has given way to other factual television formats on the main PSB channels in primetime. This is also the case for food, gardening and reality series. Travel and lifestyle series have been moved to niche (and typical 40-plus) channels, while reality series are shown on channels for young people. In addition to the growing market for their own weekly printed travel and lifestyle magazines, newspaper publishers also offer wide-ranging online travel magazines with news, tips, guidance, ads, shops and tourists' personal travel photos (Good, 2013). In the following, we will point out some trends for travel series in this media landscape.

Travel series are increasingly produced as cross-media products including websites, online shows, apps and DVD box sets. The viewer can watch the last episode online, chat with the hosts and find information about the destinations on the TV show's website. The cross-media productions change the ways of 'watching' TV since the program is omnipresent and engages viewers as users. *Pilot Guides*, for example, started as a travel series, but has extended its brand over the past decade and is now an international company producing travel journalism and entertainment for a global market. The website (www.pilotguides.com) includes more than ten TV shows with their own sub-websites offering their own destination guides and online shops for guidebooks, DVD box sets, merchandise and music. On the website users can also find recipes, follow the host and the series on social media, sign up for news-letters, share photos and videos or chat. Entertainment and consumer features are emphasized in these online and cross-media travel series. Another possibility could be travel series made for mobile devices such as smart phones and tablets. This example illustrates a radical new way of using and connecting travel series to tourist experiences, since users can watch favorite travel shows while visiting a specific destination. Locative media can even suggest travel shows that match actual loca-tion. Technological issues are still obstacles, though, such as the need for online access or the possibility to download large files.

Elements of product placement and brand content are also influenc-ing the genre. One can argue that marketing and commodification has always been an element of consumer journalism. Still, increasingly com-petitive conditions among platforms for lifestyle news and travel series, both in print and on television, have forced producers to find new ways of financing and co-producing media content. Concepts such as product or location placement and branded entertainment are popular methods for defraying production costs. Television productions include new forms of collaboration between broadcasters, production compa-nies, destinations and tourism enterprises. An example is the culinary travel series *New Scandinavian Cooking*[4] with famous new Nordic chefs showing different local specialties and menus in picturesque locations. The production is in English and meant for an international media (and tourism) market based on co-operation with the chefs, and with travel and food companies as well as with public partners and tourist boards. An example of another innovative business model is the online video travel guide *TravelGuide.TV*,[5] a web TV channel established in 2008 where tourism companies and filmmakers can show online video-clips.

Given the increasingly commercialized conditions of production, we end this chapter by asking whether these travel series can be considered journalism. As media entertainment, this kind of production has the primary goal of capturing and keeping the viewer's or reader's interest. In travel and lifestyle journalism, classical news criteria such as evidence and source assessment tend to give way to joyful stories, practical guides and pleasurable images. Instead of evaluating this type of programming with criteria established in news journalism, we suggest a broader concept of journalism that considers entertainment as valid a motivation as information in news journalism. Accordingly, journalism can cover different discursive fields. The travel series is a hybrid genre, in which a mix of different elements encompasses elements from journalism, television entertainment and marketing.

Notes

1. Danish National Research Project: TV entertainment – crossmediality and knowledge, http://tvunderholdning.au.dk.
2. These concepts were discussed at the lifestyle media conference in Brighton, May 2010. David Dunn among others took part in the conference, and suggested the three concepts.
3. One researcher functioned as primary coder, while the second tested coding independently.
4. Corporate website: http://www.newscancook.com/
5. www.travelguide.tv

References

Andersen, Michael Bruun (1994) 'Tv og genre', in Peter Dahlgren (ed.), *Den Mångtydiga Rutan*. Stockholm: Skriftserien JMK, pp. 8–37.
Bell, David and Hollows, Joanne (eds) (2005) *Ordinary Lifestyles – Popular Media, Consumption and Taste*, Glasgow: Open University Press.
Berelson, Bernard (1952) *Content Analysis in Communication Research*, Glencoe, Illinois: Free Press.
Bonner, Frances (2003) *Ordinary Television*, London: Sage.
Bruun, Hanne (2010) 'Genre and Interpretation in Production: A Theoretical Approach'. *Media Culture & Society*, 32.5, pp. 723–37.
Carlsen, John and Frandsen, Kirsten (2005) 'Nytte- og livsstilsprogrammer', Working Paper 133.05, Center for Kulturforskning: Aarhus University. Available at: http://www.hum.au.dk/ckulturf/pages/publications/jc/livsstilsprogrammer.html
Christensen, Christa Lykke (2008) 'Livsstil som tv-underholdning', *Mediekultur*, 24.45, pp. 23–36.
Christensen, Christa Lykke (2010) 'Livsstilsprogrammer', in Hanne Bruun and Kirsten Frandsen (eds), *Underholdende TV*. Aarhus: Aarhus Universitetsforlag, pp. 121–45.
Dunn, David (2005a) 'Playing the tourist', in David Bell and Joanne Hollows (eds) *Ordinary Lifestyles – Popular Media, Consumption and Taste*. Milton Keynes: Open University Press, pp. 128–142.
Dunn, David (2005b) 'We are not here to make film about Italy, we are here to make a film about ME...', in David Crouch, Rhona Jackson and Felix Thompson (eds), *The Media and the Tourist Imagination*. London: Routledge, pp. 154–169.

Eide, Elisabeth and Simonsen, Anne Hege (2004) *Å se verden fra et andets sted: medier, norskhet og fremmedhet*, Oslo: Cappelen

Frandsen, Kirsten and Bruun, Hanne (2005) 'Mediegenre, identifikation og reception', *Mediekultur*, 21.38, pp. 51–61.

From, Unni (2007) 'Forbruger- og livsstilsjournalistik – en analyse af nytte og nydelse i journalistikken', *Mediekultur*, 23.42/43, pp. 35–45.

From, Unni and Kristensen, Nete Nørgaard (2011) *Kulturjournalistik – journalistik om kultur*, København: Samfundslitteratur.

From, Unni and Waade, Anne Marit (2007) 'Smagfulde fremstillinger - Oplevelsesmatricer i mad- og rejselivsjournalistik', *Journalistica*, 4, pp. 68–92.

Fürsich, Elfriede (2002a) 'How can global journalists represent the 'Other'?', *Journalism*, 3.1, pp. 57–84.

Fürsich, Elfriede (2002b) 'Packaging Culture: The Potential and Limitation of Travel Programs on Global Television', *Communication Quarterly*, 50.2, pp. 204–26.

Good, Katie Day (2013) 'Why We Travel: Picturing Global Mobility in User-generated Travel Journalism', *Media, Culture and Society*, 35.3, pp. 295–313.

Hanitzsch, Thomas (2007) 'Deconstructing Journalism Culture: Toward a Universal Theory', Communication Theory, 17.4, pp. 367–85.

Hanusch, Folker (2010) 'The Dimensions of Travel Journalism: Exploring new fields for journalism research beyond the news', *Journalism Studies*, 11.1, pp. 68–82.

Hill, Annette (2007) *Restyling Factual TV: Audiences and news, documentary, and reality genres*, Abingdon: Routledge.

Hsieh, Hsiu-Fang, and Sarah E Shannon (2005) 'Three Approaches to Qualitative Content Analysis', *Qualitative Health Research*, 15.9, pp. 1277–88.

Jacobsen, Louise B. (2008) 'Hello My Name Is Frank Hvam: Autofictional Humour in the Danish TV series "Klovn"', *P.O.V.: A Danish Journal of Film Studies*, 26, pp. 88–98.

Jansson, André (2002) 'Spatial Phantasmagoria: The Mediatization of Tourism Experience', *European Journal of Communication*, 17.4, pp. 429–443.

Jantzen, Christian and Vetner, Mikael (2008) 'Underholdning, emotioner og identitet, et mediepsykologisk perspektiv på underholdningspræferencer', *Mediekultur*, 24.45, pp. 3–22.

Johnson, Catherine (2012) *Branding television*, Abingdon: Routledge.

Kabel, Lars (2005) *Verden langt herfra - en analyse af nyheder og faktaprogrammer i dansk tv*, Report for The Ministry of Foreign Affairs of Denmark, Center for Journalistisk Efteruddannelse, Aarhus. Available at: http://130.225.180.61/cfje/Kildebase.nsf/c62a7ee1003930acc12569570046a5c6/b318fae7982cbacec12570 05002e4bd1/$FILE/Verden%20langt%20herfra.pdf

Neale, Stephen (1980) *Genre*, London: British Film Institute.

Nichols, Bill (2001) *Introduction to Documentary*, Bloomington, IN: Indiana University Press.

Orgeret, Kristin Skaret (2002) 'Med gutta på tur; Blikk på verden i TV2's reiseprogram', in Gun Sara Enli, Trine Syvertsen and Susanne Østby Sæther (eds) *Et hjem for oss – et hjem for deg? Analyser av TV2 1992 – 2000*. Oslo: Ij-forlaget AS, pp. 182–201.

Palmer, Gareth (ed.) (2008) *Exposing Lifestyle Television: The Big Reveal*, Aldershot: Ashgate.

Rojek, Chris (2001) *Celebrity*, London: Reaktion.

Scannell, Paddy (1994) 'Kommunikativ intentionalitet i radio og fjernsyn', *Mediekultur*, 10.22, pp. 30–41.

Sørensen, Anders (1999) *Travellers in the Periphery: Backpackers and other Independent Multiple Destination Tourists in Peripheral Areas*, Nexø: Unit of Tourism Research at the Research Centre of Bornholm.

Urry, John and Larsen, Jonas (2011) *The Tourist Gaze 3.0*, London: Sage.

Waade, Anne Marit (2006). 'Armchair Traveling with Pilot Guides - Cartographic and Sensuous Strategies', in Jesper Falkheimer and André Jansson (eds) *Geographies of Communication: The Spatial Turn in Media Studies*. Gothenburg: Nordicom, pp. 155–68.

Waade, Anne Marit (2009) 'Travel Series as TV Entertainment', *Mediekultur*, 25.46, pp. 100–116.

Waade, Anne Marit (2010) 'Imaging Paradise in Ads, Imagination and Visual Matrices in Tourism and Consumer Culture', *Nordicom Review*, 31.1, pp. 15–33.

Zhang, Yan, and Wildemuth, Barbara M. (2009) 'Qualitative Analysis of Content', in Barbara M. Wildemuth (ed.) *Applications of Social Research Methods to Questions in Information and Library Science*. Westport, CT: Libraries Unlimited, pp. 308–20.

4

Framing Tourism Destination Image: Extension of Stereotypes in and by Travel Media

Steve Pan and Cathy H. C. Hsu

Introduction

While the important relationship between media and tourism has long been recognized in general (Beeton, Croy and Frost, 2006; UNWTO, 2009), less is known about the specific role media play in the creation of destination image. Besides tourism motivation, destination image is one of the most extensively researched areas in tourism studies. However, most image studies focus on tourists' perceptions of destinations, despite the fact that it has been established that a destination's image is formed through many agents (Gartner, 1993). To create a distinguishable, uniform and favorable image of a destination, for example, destination marketing organizations (DMOs) must collaborate widely with media (Beerli and Martin, 2004). Within tourism studies, mass media tend to be considered as examples of brokers of tourism (Miller and Auyong, 1998). Brokers are defined as those who 'receive monetary remuneration for an involvement with tourism production' (Cheong and Miller, 2000, p. 379), and they are important for the success of a tourism program (Cheong and Miller, 2000).

The interest in destination image is based on the idea that intermediaries' (brokers') images and knowledge of a destination will arguably have a significant impact on potential tourists' vacation decision-making processes. Citrinot (2007) argued that to 'market a destination is all about discovering ways of generating dreams for potential tourists', and DMOs need to decide on promotional content and channels in order to create and deliver those dreams to target markets.

The importance of a positive image for travel promoters was confirmed by Baloglu and Mangaloglu (2001), who investigated tourism destination images of Turkey, Egypt, Greece and Italy, as perceived by

US-based tour operators and travel agents. Their findings indicate that travel intermediaries were unlikely to promote a destination of which they had a negative or weak image. Yet, the question remains how the media factor in this process of image creation.

While the variety of stakeholders in the process of image creation sometimes has confounded the study of destination image, in this chapter we hope to shed light on some of this complexity in relation to the media by combining approaches from tourism studies with concepts from journalism studies. Especially the concept of framing is used to elucidate how media engage in destination image creation. This chapter aims to illuminate factors that affect travel media content, which may reflect and carry expectations and stereotypes. We begin with a review of the literature and, by linking tourism and journalism studies, we put forward two propositions about the coverage of destinations in travel media. We then move on to test the proposed relationships with actual research and data analysis by examining travel magazines articles.

Tourism and media: image and framing

Destination image formation processes in tourism are equivalent to framing theory in mass communication. The creation of destination images through travel brochures, TV commercials and other channels involves selection, exclusion and emphasis of certain images (Wang, 2000). According to Entman (1993, p. 52), to frame is 'to select some aspects of a perceived reality and make them more salient in a communication text'. The purpose of framing is to influence audiences' interpretations of an event through selection, exclusion and emphasis (Entman, 1993; Gitlin, 1980; Tankard, 2003). Destination image can be defined as 'the expression of all objective knowledge, impressions, prejudices, imaginations, and emotional thoughts with which a person or group judges a particular object or place' (Lawson and Baud-Bovy, 1977, p. 10). This definition implies to a certain extent typecasting or stereotyping. On the other hand, image is a mental construct based on selected few impressions among the flood of total impressions (Fakeye and Crompton, 1991; Reynolds, 1965). Destination image, therefore, is the product of a framing process.

By repeating and highlighting certain keywords, concepts, symbols and images, but not others, journalists create news with texts and images that contain certain frames through which audiences 'see' a news story (Entman, 1991; Pan and Kosicki, 1993). The more the news frames are congruent with audiences' frames, the more likely they will influence audiences' opinions on specific issues. By the same token, travel journalists create stories that contain frames of reference (keywords and images) through which their audiences construct mental images of a destination. As in communication, where public opinion can change after media frames change (Lang and Lang, 1983), destination image may be altered after a change in (travel) media frames. The change of frames can be detected through chronological analysis of media content and/or a survey of the general public.

A similar concept to framing has been proposed by geographical scholars in studying mental maps of a place. Gould and White (1974) argue that many of the human patterns we see on the landscape today are the result of people making spatial decisions based on information that has come through a perceptual filter, or frame. Human behavior is therefore affected only by that portion of the environment that is actually perceived. Tourists also act upon their perception, rather than reality (Guthrie and Gale, 1991). We cannot absorb and retain the virtually infinite amount of information that impinges upon us daily. Actually, we see more than 3,156 images a day, and only about 30 reach our consciousness (Creswell, 2008). Thus, we apply perceptual filters to screen out most information in a highly selective fashion (Gould and White, 1974). In mass communication, these perceptual filters can be molded through framing, and the same logic applies to tourism destination image formation.

Applying frame analysis to travel journalism can help researchers better understand the power struggle (or interactions) (Entman, 1991) between different frames' sponsors and adopters. This is important for our understanding of what factors shape travel journalism content, as most of the extant studies on destination image focus on image perception from visitors.

Media images of destinations

A place's image is framed in media in a fundamental way through news values. This is because news values motivate the selection and presentation of news by journalists and editors (Gamson, 1992; Price and Tewksbury, 1997), which in turn creates the frames in the news. It is generally agreed that there are five news values that are most likely to trigger an event to be covered in the news: conflict, drama, personalization, proximity (closeness to home) and novelty (Price and Tewksbury, 1997). Galtung and Ruge (1973) used the term 'meaningfulness' instead of proximity. This term also entails relevance, ethnocentricity and cultural proximity.

In travel journalism, 'conflict' tends to be associated more with psychological contrast than with the physical conflict commonly found in political news. Journalists (and visitors) are seeking something old (conventional and/or familiar), something new (contemporary and/or different), something indigenous and something universal. 'Drama' is related to the dramatic or dynamic nature of a destination's attractions and activities. A well-known example is the Festival of San Fermín (otherwise known as 'the running of the bulls') in Pamplona, Northern Spain, made famous by Ernest Hemingway's 1926 novel *The Sun Also Rises*. 'Personalization' is more closely related to descriptions of customized trips or special-interest tourism. 'Proximity' is mainly concerned with the establishment of physical and cultural distance. 'Novelty' is a distinctive motivation for traveling as the pursuit of difference; travel journalism can echo this escape from the everyday life circumstances. These news values are interwoven with leisure motivations (Ryan, 1995), especially those of a social (personalization) and intellectual (novelty and drama) nature. In sum, travel journalists tend to report on destinations that are different, contrasting, dramatic and novel to their target readers.

Images of places in news media can be categorized into rich or one-dimensional images (Avraham, 2000). It is generally accepted that places that consolidate political, economic, cultural and technological powers usually attract more media coverage in terms of news variety and quantity (Wu, 2000). On the contrary, places with geographically, politically and economically peripheral positions usually have a one-dimensional image. They are either ignored by news media coverage or covered in a uniform way. When there is not enough news coverage on a certain place, in terms of quantity and variety, audiences arguably do not have enough information to construct a comprehensive and objective mental image of that place. That is, low familiarity can potentially lead to a less positive image.

The mental map of a place

How we evaluate places is partly influenced by the amount and type of information we have about different localities: local scenery (landscape), climate, cultural and linguistic variety, quality of education, political and social attitudes and accessibility (relative location) (Gould and White, 1974). These aspects – proposed by scholars of geography – are very similar to those proposed by tourism scholars. For example, Beerli and Martìn (2004) proposed nine perceived destination image dimensions/attributes: natural resources, general infrastructure, tourist infrastructure, tourist leisure and recreation, culture, history and art, political and economic factors, natural environment, social environment and atmosphere of the place. A person's mental map of a place is a subtle mix of a shared, national viewpoint (or a widespread common sense), and local desirability, representing feelings people have for the familiar and comfortable surroundings of their home area (Gould and White, 1974). Kaplan (1973) has postulated that an enormous amount of information is stored in the form of cognitive maps with distinct spatial properties such as location, distance and direction. Theoretically, it is generally accepted that the cognitive component is an antecedent of the affective component, and knowledge of the objects is a foundation of the subjects' evaluative responses (Beerli and Martin, 2004; Sönmez and Sirakaya, 2002).

In Baloglu and McCleary's (1999) study, images (for both tourists and non-tourists) of four Mediterranean destinations (Turkey, Egypt, Greece and Italy) were compared in cognitive, affective and overall terms, for the purpose of understanding their relative strengths and weaknesses. Baloglu and Brinberg (1997) evaluated 11 Mediterranean countries as tourism destinations on each of Russell and Pratt's (1980) two orthogonal bipolar scales of pleasant–unpleasant and arousing–sleepy or exciting–gloomy and relaxing–distressing. Their findings indicate that the affective images of tourism destination countries varied across both positive and negative dimensions. This study presents an example of affective destination image positioning and shows that destinations can differentiate themselves from competitors based on different affective attributes.

Increasing competition among international tourist destinations has highlighted the importance of destination positioning. As Carter (2003, pp. iii79) explained, 'the position of a product was a place in consumers' minds, but it was a place relative to competing products'. Since it is generally agreed that creating and transmitting a favorable image is the core or key construct of positioning strategy (Baloglu and Mangaloglu, 2001; Baloglu and McCleary, 1999; Echtner and Ritchie, 1991; Pike and Ryan, 2004; Sönmez and Sirakaya, 2002), the investigation of a destination's image in relation to competitors' images should provide more insight into the positioning strategy formulation than the study of a single destination could have achieved. Guided by this rationale, many studies (Baloglu and Brinberg, 1997; Baloglu and McCleary, 1999) with an extensive review of research on destination image and positioning (Pike and Ryan, 2004) explore the images of multiple destinations with the positioning implication in mind. Similar to comprehensive and reliable frames that need to be examined through comparison of narratives (Entman, 1991), destination (or place) image also needs to be examined through comparisons with other places (Gould and White, 1974; Strauss, 1961).

Having outlined the relationship between media and tourism, the remainder of this chapter will be devoted to an analysis of media content in order to identify the destination image projected through print media, which can be interpreted as a reflection and a proxy of the perceived images of their readership. Both of the images projected by media and perceived by the readership are to a certain extent conditioned by the stereotypes the mainstream society has for these destinations.

The goal of this empirical analysis is to enhance further our understanding of the image formation process during this process. Through the analysis of travel magazines, this chapter tries to offer answers to explain the production of geographical stereotypes and the relationship between news media image and travel media image. Like other media channels, travel magazines shape their reports (travelogues) to elicit favorable reactions from their readership; moreover, this coverage is also influenced by public relations and promotion campaigns of DMOs who are the primary 'sponsors' and 'suppliers' of travelogues' frames. These frames act as filters for the readers to facilitate their processing of travel information to form a destination image. Our two propositions are:

Proposition 1: Media tend to re-present geographical stereotypes. Destinations' locations on travel media's image-perceptual map resemble their locations on the physical map.
Proposition 2: Places with rich news media images are likely to have rich travel media images. Travel journalists report various aspects of a destination when it has rich news media images.

Methodology

To test the two propositions, travel media in China were selected for analysis. China's outbound tourists have increased by more than 400 per cent since 2002 to reach 70 million in 2011 (CNTA, 2012), and it has become Asia's largest tourist-generating country (International Forum on Chinese Outbound Tourism, 2009). It is important to understand how travel media in China see the outbound destinations, because travel media images of these places help provide frames of reference for readers, especially when they have not visited the destinations before. The top three outbound destinations for Mainland Chinese in the timeframe we analyzed (2006–2008) were Hong Kong, Macau and Japan, with Korea and Vietnam vying for the next two spots (CNTA, 2007, 2008, 2009). These top five destinations were selected for analysis. As Taiwan had gradually relaxed its travel restrictions on Mainland Chinese tourists, it was also included in the final analysis due to its potential for becoming a major outbound destination.

In order to test the propositions, articles published in China's top six travel magazines (based on circulation during the period of 2006–2008) were collected. The magazines – *Voyage, National Geographic Traveler, Traveler, World Travel Magazine, Traveling Scope* and *Travel + Leisure* – all enjoyed a monthly circulation of 300,000 or more (MediaSearch.cn, 2009). The first three have headquarters in Beijing while the other three are based in Shanghai. Most of these publications cooperated closely with international travel magazine publishers and adopted the same foreign titles such as *National Geographic Traveler* and *Travel + Leisure* with Chinese translations. Altogether, 413 articles covering the six destinations were published during this period. The articles were digitized using optical character recognition software and crosschecked for accuracy by two research assistants.

Qualitative analysis software Nvivo 8 was used to analyze the Chinese textual data. A list of word frequency was generated by the software. The authors then checked the top 100 most frequently appearing Chinese characters and ignored some conjunctions like 'and' or 'or'. The authors also checked the original context that contained the selected characters to ensure they were meaningful to be included for further analysis. In English, words are composed of letters of the alphabet and are used as basic units to express meaning. In Chinese, characters alone have their own meaning and in combination with other characters they form terms to express the same or other meanings. Finally, a list of the frequently appearing Chinese terms (or keywords) and expressions was prepared. The data set was mined with this list, and the resulting frequencies of these terms or expressions were recorded.

Due to the enormous number of Chinese terms and expressions, the authors grouped terms with a similar meaning into the same category. For example, seafood, fish and shrimp were combined into 'seafood'; Mickey Mouse and Hello Kitty were combined as 'cartoon characters'; cold noodles, noodles, tart, spring roll and stone-grilled rice were combined as 'specialty food' and so on. All of the selected Chinese terms or expressions were grouped into nine categories (destination image dimensions), as proposed by Beerli and Martin (2004). However, there were no keywords in the analyzed articles that were associated with the dimension 'Social Environment', reducing to eight the number of final dimension. Moreover, two minor changes were made. First, the component 'beauty of the scenery' in the dimension 'Natural Environment' was transferred to 'Natural Resources', which includes eight attributes of scenery, hot spring, ocean (coast and beach), and volcano. Second, the dimension 'Natural Environment' was renamed 'Physical Environment' since most of the components were actually describing the physical condition of a destination, such as constructions, fishing village and public housing.

Correspondence Analyses (CA) were conducted to test Proposition 1. CA is a perceptual mapping technique designed to explore and describe the correspondence between row and column variables (Greenacre, 1993). It is a method of data analysis for representing tabular data graphically and is suitable to test Proposition 1, as the main purpose is to map perceptually the correspondences between framed image attributes and destinations. CA plots of the row profiles (relative frequencies) were constructed so that each column category became a different dimension. When there are more than three column categories, researchers have to project the high dimensional space onto a two-dimensional subspace to be visually interpretable and maintain as much of the original information (or variation) as possible. In the current study, the row variables are six destinations and the column categories are eight image attributes for Proposition 1.

As for Proposition 2, a new Media Image Richness Index (MIRI) based on previously defined variety and frequency was constructed to measure the richness of each destination's image in travel media. Richness is measured by the number of reports (quantity) and the number of image attributes mentioned (variety). The equation to calculate the Media Image Richness Index is:

$$\text{MIRI} = \underbrace{\text{TNRD} \times \text{TNT}}_{\text{(I)}} \times \underbrace{\text{TIDC} \times \text{TID}}_{\text{(II)}}$$

TNRD: Total number of reports for each destination
TNT : Total number of terms mentioned
TIDC : Total image dimension component mentioned (%)
TID : Total image dimension mentioned (%)

The total image dimension component (TIDC) mentioned equals the number of image dimension components mentioned for individual destinations divided by the sum of total image components mentioned for the six destinations, which was 118. The total image dimension (TID) mentioned equals the number of image dimensions mentioned for individual destinations divided by the total number of image dimensions, which was eight. Part I of the equation calculates 'quantity', while Part II measures 'variety' of travel reports. The higher the MIRI, the richer the destination image described in travel media. MIRI was further transformed into a scale ranging from zero to 100. We first deducted the lowest value from all values so that the new lower limit is zero. We then multiplied all values by a quotient of new highest value (which is 100) divided by the old highest value (after deducting the old lowest value) so that the new higher limit was equal to 100.

Analysis

Frequencies of terms or expressions denoting meanings of components of the eight image dimensions are tabulated in Table 4.1. For Japan, for example, terms related to 'leisure and recreation' and 'culture, history and art' were mentioned 2,460 and 1,852 times respectively by the travel magazines, while no travel journalists mentioned 'atmosphere of the place'. It was also noted that Japan was the only destination with which the dimension 'political and economic factors' was associated. This reflects the historical legacy and power struggle between China and Japan, and this mentality revealed in Chinese travel magazines is particular to Japan. These frames are the product of interactions (or 'power struggle') among the tourist-generating society (stereotypes), destination marketing organizations (promotional and public relations activities), editorial policies and readers' preferences. However, to understand the relationships between image dimensions and destinations requires the display of a comprehensive perceptual map.

Proposition 1 argued that media tend to re-present geographical stereotypes. Destinations' locations on the travel media images perceptual map bear resemblance to their locations on the physical map. As 'political and economic factors' is an outlier, this image dimension is treated as a supplementary (passive) point, as suggested by Greenacre (2007), so that interesting contrasts between the more frequently occurring categories are not completely masked. CA results indicate that a five-dimension solution explains 100 % of the variation. However, to be interpretable and visually presentable, a two-dimensional solution (see Table 4.2) which explains 90 per cent of the variation, is deemed adequate for further analysis as the eigenvalues of these two dimensions

Table 4.1 Distribution of terms denoting perceived destination image

Image Attributes (components*)	Japan	Korea	Taiwan	Hong Kong	Macau	Vietnam
1. Natural resources (8)	1,001	379	269	35	0	27
2. General infrastructure (9)	36	0	93	104	18	136
3. Tourist infrastructure (12)	413	38	255	685	304	342
4. Leisure and recreation (17)	2,460	895	1,102	1,447	354	651
5. Culture, history and art (63)	1,852	1,312	743	930	342	665
6. Political and economic factors (2)	32	0	0	0	0	0
7. Physical environment (4)	178	0	46	56	0	27
8. Atmosphere of the place (3)	0	89	0	24	8	29

* number of items included under their corresponding image dimensions.

Table 4.2 Results of correspondence analysis

Dimension No.	Eigenvalue (Singular Value)	Inertia	% Explained	Cumulative %
1	0.3360	0.1112	67.9	67.9
2	0.1970	0.0374	22.7	90.6

Object	Contribution to Total Inertia (%)	Contribution to Inertia (‰)		Explanation by Dimension (‰)	
		Dimension I	Dimension II	Dimension I	Dimension II
Attributes:					
Natural resources	293	401	74	930	58
General infrastructure	133	110	1	559	1
Tourist infrastructure	332	460	15	941	10
Leisure and recreation	28	1	89	18	711
Culture, history and art	79	23	276	199	790
Political and economic factors	0	0	0	930	58
Physical environment	41	3	148	50	815
Atmosphere of the place	93	2	398	17	967

Raw Counts Converted into Chi-squared Contributions

Destinations:

	Japan	South Korea	Taiwan	Hong Kong	Macau	Vietnam
Japan	209	215	266	699		289
Korea	277	206	597	505		490
Taiwan	27	0	52	0		442
Hong Kong	191	265	5	939		6
Macau	143	159	12	758		19
Vietnam	154	155	68	685		101

	Japan	South Korea	Taiwan	Hong Kong	Macau	Vietnam
Natural resources	0.1292	0.0519	0.0106	-0.1216	-0.0762	-0.0881
General infrastructure	-0.0638	-0.0590	0.0377	0.0274	-0.0077	0.1105
Tourist infrastructure	-0.0823	-0.1191	-0.0173	0.1162	0.1271	0.0624
Leisure and recreation	0.0133	-0.0424	0.0252	0.0299	-0.0203	-0.0265
Culture, history and art	-0.0265	0.1004	-0.0262	-0.0396	-0.0012	0.0102
Political and economic Factors	0.0480	-0.0170	-0.0163	-0.0186	-0.0104	-0.0141
Physical environment	0.0535	-0.0525	0.0019	-0.0020	-0.0323	-0.0081
Atmosphere of the place	-0.0545	0.1028	-0.0353	-0.0062	-0.0022	0.0241

are either higher than or close to 0.2 (Hair et al., 1998). This trade-off between interpretability and variation explanation makes some points not well represented in the conceptual map.

We can try to determine what aspect of image attribute and destination has the strongest relationship (or 'correspondence') by looking at each component's (row and column) contribution to the overall chi-square statistics (see Table 4.2). The statistics measure the differences between row profile (relative frequency) points and their average; the higher the statistic, the more variation the row profiles have. A higher variation means the row profiles are more dispersed and are closer to the column profiles (Greenacre, 2007). Therefore, the higher the contribution, the stronger the relationship and the closer the row and column profiles should be located on the perceptual map (Figure 4.1) (Hair et al., 1998).

For example, the high values of Japan and Korea with the image dimension 'natural resources' indicate a stronger relationship between the destinations and the dimension, and they therefore are placed close together on the perceptual map. Japan has a more salient travel media image in natural resources, physical environment (architecture) and politics. Korea has a more prominent image in culture, history and art, the ambience and natural resources. It is noted that the atmosphere associated with Korea is not so much contemporary as conventional. Taiwan is represented in travel media as a place with urban landscape (leisure and recreation) and ease of currency exchange. Hong Kong has a traditionally strong image in tourist infrastructure and is a place full of leisure and recreation opportunities. Macau, restricted by its small land area, has a one-dimensional travel media image, which is rich only in tourist infrastructure with a focus on hotels. This explains why the Macau government in recent years has been trying hard to diversify tourist attractions in order to enrich its destination image and diversify its market segments. Vietnam, an emerging destination, has a strong travel media image in general infrastructure, atmosphere and culture, history and art. The rail link between Yunnan Province, China and Vietnam contributes strongly to the general infrastructure image, while French and Chinese influences and dining (especially coffee drinks) are the most frequently mentioned components of cultural image components.

The inertia is the weighted average of the squared χ^2-distances between the profiles and their average profile. The smaller the inertia, the less dispersed the profiles are in their profile space. As there are eight attributes, the threshold for determining the level of contributions to

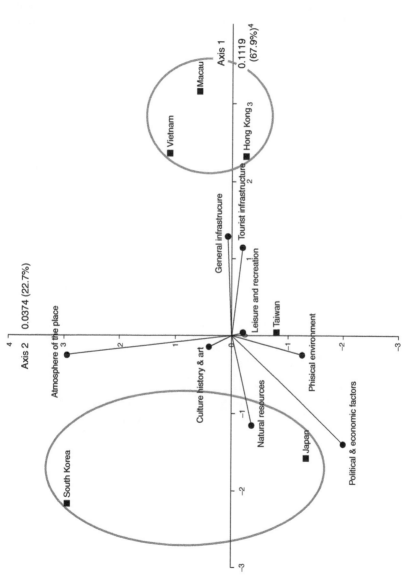

Figure 4.1 Perceptual map of media image dimensions and destinations

total inertia is 125‰ (1000/8). The major contributors to the variance are natural resources, general infrastructure and tourist infrastructure. For destinations, the main contributors to the variance are Japan, Korea and Hong Kong. As for dimension I (axis I), its variance is mainly explained by image dimensions of natural resource and tourist infrastructure, and by destinations of Japan, Korea and Hong Kong. The further we move toward the right side of axis I, the more man-made infrastructure and the more south or the lower latitude the destination is. By rotating Figure 4.1 clockwise 90 degrees and superimposing it on the physical map, we can produce Figure 4.2. This travel media image map largely corresponds to the geographical location of each destination defined by its latitude and longitude. The image dimension of 'leisure and recreation' is not displayed in Figure 4.2 as it is almost universally present (hence a low contribution to inertia). Looking at Figure 4.2, we can say that the Chinese travel media image of the six destinations can be described as 'northeast with nature and culture' and 'south with infrastructure'.

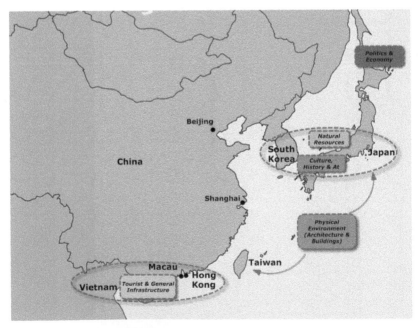

Figure 4.2   Destinations' travel media images on physical map

Destination image benefits from a rich media image

Proposition 2 argued that places that have rich news media images are likely to have rich travel media images. It was expected that travel journalists report various aspects of a destination when a destination has an overall rich media image. We cannot ignore the fact that 'important news' of politics, economy, culture, technology and entertainment come mainly from certain countries/regions in the study sample. For example, Hong Kong is a global financial center and has one of the largest film industries in the world. Japan is a prominent player in the global economy, politics, technology and popular cultures. Korea has been gaining prominence in technology and entertainment development in the past two decades, since the hosting of the 1988 Seoul Olympic Games. Moreover, Japan and Korea both have global TV channels (NHK World and Arirang TV), which increase their media exposure globally. These destinations tend to have a richer place image in terms of news topics diversity and quantity (Avraham, 2000).

Out of 413 articles published by the six travel magazines during the period of 2006 to 2008, Japan is the most frequently covered country (31 per cent), followed by Hong Kong (25 per cent) and Korea (16 per cent). When we combine the quantity (number of reports) and variety (number of image dimensions and their components) to calculate MIRI, the top three destinations are Japan, Hong Kong and Korea (Table 4.3). Proposition 2 is therefore supported, as we found that when a destination has a richer image in traditional news media, its travel media image also tends to be richer.

Discussion and conclusion

The travel media perceptual map based on image attributes reconfirms the stereotypical geographical location of each destination. The further we move northeast (Korea and Japan), the more natural, cultural and historical attributes were reported on and associated with the destinations by the travel magazines. Taiwan, with its unique geographical location and historical background for Mainland Chinese tourists, has its own travel media image that is salient in leisure and recreation, physical environment and general infrastructure. Hong Kong, Macau and Vietnam, on the other hand, were clustered closely on the other side of the perceptual map. Hong Kong and Macau, two places traditionally closely attached to Canton Province of China and the top two visited destinations by Mainland Chinese, both were reported more in terms of their tourist infrastructure, as a place to play. Vietnam, from

Table 4.3 Variety and quantity of travel magazines reports

	Japan	Korea	Taiwan	Hong Kong	Macau	Vietnam
Total number of reports (a)	129	67	49	104	26	38
Total number of terms (keywords) mentioned (b)	11,944	5,426	5,126	6,562	2,052	3,798
Total image attributes components mentioned (c)	51	35	41	35	26	36
Total image attributes components mentioned (%) (d)	43	30	35	30	22	31
Total image attributes mentioned (e)	8	6	7	8	6	8
Total image attributes mentioned (%) (f)	100	75	88	100	75	100
Media Image Richness Index (a×b×d×f)	665,929	80,873	76,363	202,421	8,817	44,031
Transformed MIRI	100	11	10	29	0	5

Item (d) equals individual number of (item c) divided by total number of image components, i.e., 118.

Item (f) equals individual number of (item e) divided by total number of image attributes, i.e., 8.

a historical point of view, is close to southern China both in physical and cultural distance. This historical stereotype is still reflected in the travel media's perceptual map. Korea and Vietnam both are more salient in the image dimension 'atmosphere of a place'. They possess an advantage over other destinations in terms of affective image, which, according to Baloglu and Brinberg (1997), is a main factor in differentiating one destination from others. In this study, 'atmosphere of a place' is comprised of novelty, fashion and tradition. Korea is prominent in 'tradition' as reported by travel journalists, while Vietnam is a blend of fashion and tradition (Santos, 2004).

The support for Propositions 1 and 2 indicates that travel journalists seldom challenge, but rather reflect the dominant frames in their own culture. These dominant frames, if they persist long enough, become stereotypes and constitute the mental pictures for readers to understand the destinations. Travel journalists report what they assume their audiences would like to read, see and know. By doing so, they participate in the process of extending and reinforcing the stereotypes which, in turn, constrain and condition the reported frames. Other than the stereotypical frames, the media's editorial policies can strongly affect journalists' reported frames, and these policies are also subject to influences from DMOs through public relations campaigns (familiarization trips), as well as from advertisers. Hence, the reported travel media frames are a composite of different stakeholders in destination image. In the future, travel media frames studies could also include the study of influences imposed by advertisers and advertising agencies.

Japan, Hong Kong and Korea are traditionally richer in international news media image due to their political, economic, cultural, technological and geographical importance. This richness in news media image translates into rich travel media image in terms of quantity and variety of reports. These findings demonstrate that tourism generally benefits from a rich, favorable and strong country image, which is an amalgam of political, socio-cultural, economic, technological and environmental development over time. Because there are many sectors involved, it explains the findings in this chapter that tourism destination image as carried in the travel media is stereotypical and slow, if not hard, to change. In the future, the study of tourism destination image needs to take a multi-disciplinary approach that includes advertising, journalism, economic development, cultural values, geography and history in order to chronologically and comprehensively trace the changes of destination image.

References

Avraham, Eli (2000) 'Cities and Their News Media Images', *Cities*, 17.5, pp. 363–70.

Baloglu, Seyhmus and Brinberg, David (1997) 'Affective Images of Tourism Destinations', *Journal of Travel Research*, 35.4, pp. 11–15.

Baloglu, Seyhmus and Mangaloglu, Mehmet (2001) 'Tourism Destination Images of Turkey, Egypt, Greece, and Italy as Perceived by US-based Tour Operators and Travel Agents'. *Tourism Management*, 22.1, pp. 1–9.

Baloglu, Seyhmus and McCleary, Ken W. (1999) 'U.S. International Pleasure Travelers' Images of Four Mediterranean Destinations: A Comparison of Visitors and Nonvisitors', *Journal of Travel Research*, 38.2, pp. 144–52.

Beerli, Asunciòn and Martin, Josefa D. (2004) 'Factors Influencing Destination Image', *Annals of Tourism Research*, 31.3, pp. 657–81.

Beeton, Sue, Croy, Glen and Frost, Warwick (2006) 'Tourism and Media into the 21st Century', *Tourism, Culture & Communication*, 6.3, pp. 157–9.

Carter, Stacy (2003) 'The Australian Cigarette Brand as Product, Person, and Symbol', *Tobacco Control*, 12 (Supplement III), pp. iii79–iii86.

Cheong, So-Min and Miller, Marc L. (2000) 'Power and Tourism: A Foucauldian Observation', *Annals of Tourism Research*, 27.2, pp. 371–90.

Citrinot, L. (2007) 'It's All About Image and Money', *TravelWeekly*. Available at: http://www.travelweeklyweb.com/article/it_is_all_about_image_and_money.html

CNTA. (2007) *Statistic Report of China's Tourism Industry in 2006*. Available at: http://www.cnta.gov.cn/html/2008-6/2008-6-2-14-52-59-213.html

CNTA. (2008) *Statistic Report of China's Tourism Industry in 2007*. Available at: http://www.cnta.gov.cn/html/2008-9/2008-9-10-11-35-98624.html

CNTA. (2009) *Statistic Report of China's Tourism Industry in 2008*. Available at: http://www.cnta.gov.cn/html/2009-9/2009-9-28-9-30-78465.html

CNTA. (2012) *China Tourism Statistics Bulletin 2011*. Available at: http://www.cnta.gov.cn/html/2012-10/2012-10-25-9-0-71726.html

Creswell, J. (2008) 'Nothing Sells Like Celebrity', *New York Times*. Available at: http://www.nytimes.com/2008/06/22/business/media/22celeb.html

Echtner, Charlotte M. and Ritchie, J.R. Brent (1991) 'The Meaning and Measurement of Destination Image', *The Journal of Tourism Studies*, 2.2, pp. 2–12.

Entman, Robert M. (1991) 'Framing U.S. Coverage of International News: Contrasts in Narratives of the KAL and Iran Air Incidents', *Journal of Communication*, 41.4, pp. 6–27.

Entman, Robert M. (1993) 'Framing: Toward Clarification of a Fractured Paradigm', *Journal of Communication*, 43.4, pp. 51–8.

Fakeye, Paul C. and Crompton, John L. (1991) 'Image Differences Between Prospective, First-Time, and Repeat Visitors to the Lower Rio Grande Valley', *Journal of Travel Research*, 30.2, pp. 10–16.

Galtung, Johan and Ruge, Mari (1973) 'Structuring and Selecting News', in Stanley Cohen and Jock Young (eds), *The Manufacture of News: Social Problems, Deviance and the Mass Media*, London: Constable, pp. 62–72.

Gamson, William A. (1992) *Talking Politics*, NY: Cambridge University Press.

Gartner, William C. (1993) 'Image Formation Process', *Journal of Travel & Tourism Marketing*, 2.2/3, pp. 191–215.

Gitlin, Todd (1980) *The Whole World Is Watching: Mass Media in the Making and Unmaking of the New Left*, Berkeley: University of California Press.

Gould, Peter and White, Rodney (1974) *Mental Maps*, NY: Penguin.

Greenacre, Michael (1993) *Correspondence Analysis in Practice*, New York: Academic Press.

Greenacre, Michael (2007) *Correspondence Analysis in Practice*, 2nd edn., New York: Chapman & Hall/CRC Interdisciplinary Statistics.

Guthrie, John and Gale, Peter (1991) 'Positioning Ski Areas - A Case Study: Central Otago, New Zealand', in *Proceedings of the New Horizons in Tourism Conference*, Calgary, Canada, pp. 553–69.

Hair, Joseph F., Anderson, Rolph E., Tatham, Ronald L. and Black, Bill (1998) *Multivariate Data Analysis*, 5th edn., Upper Saddle River, NJ: Prentice Hall.

International Forum on Chinese Outbound Tourism (2009) *Overview*. Available at: http://www.outbound-tourism.cn/english/intro.asp

Kaplan, Stephen (1973) 'Cognitive Maps in Perception and Thought', in Roger M. Downs and David Stea (eds), *Image & Environment: Cognitive Mapping and Spatial Behavior*. Chicago: Aldine, pp. 63–78.

Lang, Gladys E. and Lang, Kurt (1983) *The Battle for Public Opinion: The President, The Press, and The Polls during Watergate*, New York: Columbia University Press.

Lawson, Fred, and Baud-Bovy, Manuel. (1977). *Tourism and Recreational Development*, London: Architectural Press.

MediaSearch.cn (2009) *Media Information on Travel Magazines*. Available at: http://www.meihua.info/mediasearch/medium,4a9b7eea-0a0d-4329-b0c8-a41e21641d8b.htm

Miller, Marc L. and Auyong, Jan (1998) 'Remarks on Tourism Terminologies: Anti-tourism, Mass Tourism, and Alternative Tourism', in *Proceedings of the 1996 World Congress on Coastal and Marine Tourism: Experiences in Management and Development*. Washington Sea Grant Program and the School of Marine Affairs, University of Washington and Oregon Sea Grant College Program, Oregon State University, pp. 1–24.

Pan, Zhongdang and Kosicki, Gerald M. (1993) 'Framing Analysis: An Approach to News Discourse', *Political Communication*, 10.1, pp. 55–75.

Pike, Steven and Ryan, Chris (2004) 'Destination Positioning Analysis through a Comparison of Cognitive, Affective, and Conative Perceptions', *Journal of Travel Research*, 42.4, pp. 333–42.

Price, Vincent and Tewksbury, David (1997) 'News Values and Public Opinion: A Theoretical Account of Media Priming and Framing', in George Barnett and Franklin J. Boster (eds), *Progress in the Communication Sciences: Advances in Persuasion*. Greenwich, CT: Ablex, pp. 173–212.

Reynolds, William H. (1965) 'The Role of the Consumer in Image Building', *California Management Review*, 7.3, pp. 69–76.

Russell, James A. and Pratt, Geraldine (1980) 'A Description of the Affective Quality Attributed to Environments', *Journal of Personality and Social Psychology*, 38.2, pp. 311–22.

Ryan, Chris (1995) *Researching Tourist Satisfaction: Issues, Concepts, Problems*, London: Routledge.

Santos, Carla (2004) 'Framing Portugal: representational dynamics', *Annals of Tourism Research*, 31.1, pp. 122–138.

Sönmez, Sevil and Sirakaya, Ercan (2002) 'A distorted Destination Image? The Case of Turkey', *Journal of Travel Research*, 41.2, pp. 185–96.

Strauss, Anselm L. (1961) *Image of the American City*, NY: The Free Press.

Tankard, Jr., James W. (2003) 'The Empirical Approach to the Study of Media Framing', in Stephen D. Reese, Oscar H. Gandy, Jr. and August E. Grant (eds.), *Framing Public Life: Perspectives on Media and Our Understanding of the Social World*. Mahwah, NJ: Lawrence Erlbaum, pp. 95–106.

UNWTO (2009) TOURCOM. Available at: http://www.unwto.org/tourcom/prev/en/prev.php?op=2.

Wang, Ning (2000) *Tourism and Modernity: A sociological Analysis*. Oxford: Pergamon.

Wu, Denis (2000) 'Systemic Determinants of International News Coverage: A Comparison of 38 Countries', *Journal of Communication*, 50.2, pp. 110–30.

Part II
Exploring the Producers: Professional Expectations, Routines and Markets

5
Travel Journalism in Flux: New Practices in the Blogosphere

Bryan Pirolli

Introduction

With nearly half of all travel bookings now taking place online in the US and UK, traditional travel agents are no longer the gatekeepers to the world's most exotic destinations. At least in most western societies, booking sites like Expedia and Priceline have eased growth, following a boom over the past 15 years or so, settling into their roles as top players in the market (WWWMetrics, 2014). The novelty of online travel planning has worn off and now we are left to make sense of the surplus of information and offers scattered online.

As tourists are increasingly looking to the Internet to plan and organize their trips, the information they consult before pressing 'confirm' on any reservation page is also increasingly digitalized. Gone are the days where the Sunday newspaper travel section or Arthur Frommer were the sole purveyors of what's worthwhile to see and do from London to Shanghai. Travel journalism is facing an identity crisis as its gatekeeping role, much like that of travel agents, is slowly slipping into the hands of those they originally sought to serve.

This chapter explores how travel journalism is evolving in the online world, especially in travel blogs. In particular, I evaluate the relationship of online travel information to journalistic standards and practices based on research of a travel blogosphere addressing Paris. The study explores both the content found on travel blogs as well as the interpretations and expectations of users who read them.

Travel online journalism

With no limits on publication space or deadlines to send copy to the printer, the public now has the chance to self-publish for mass audiences through sites like Wordpress and Blogger. Professional journalists no longer have the exclusive right to decide what information enters the public arena. The gatekeepers and those who they guard are in much more direct contact since the audience now generates content either on their own sites or through comments and forums on media sites, creating a circular exchange between the public and writers (Barzilai-Nahon, 2008). Axel Bruns (2003) posits that the gatekeeping of professionals has evolved instead into a gate*watching* role, with online bloggers and tweeters deciding what to highlight, curating their own information mosaics from the ever-expanding source of blogs, news sites, participatory sites and social media.

Niche media and specialized journalists, especially in the umbrella category of lifestyle journalism, are also adapting to expansive user-generated content online (Duffy and Ashley, 2012; Hanusch, 2012; Rocamora, 2012). Travel journalism is a marked example of how the fundamental role of the journalist has nearly been usurped by online players. Moreover, these journalists face stiff competition from free review sites like TripAdvisor and thousands of independently edited blogs that provide similar information in a seemingly more efficient manner.

The evolution of travel journalism

Travel writing has evolved from simple descriptions of Greek travelers to medieval travel logs, to explorers' quasi-scientific journals and to 19th-century narrations of Flaubert's exploits in Egypt (Blanton, 1997; Marcil, 2007). Few of these writings were 'journalism' in the sense of today's craft, driven by objectivity, official sources or standard formats. On the contrary, the writers were often scientists or pleasure travelers who were simply recounting what they experienced, with opinions, emotions and judgments alongside factual descriptions. Often directly linked to advances in technology, with the development of the steamboat, railroad, telegraph and commercial airlines, travel writing has been evolving over generations, a practice constantly in flux.

While more personal writings were the *ordre du jour* throughout much of the 17th and 18th centuries, the purpose for writing about travel changed as more people were able to assume the role of tourist following the Industrial Revolution, the expansion of railways, and, eventually, automobile travel. The more objective travel guide, offering a more journalistic look at crisscrossing the world and devoid, at least mostly, of the author's own private commentaries, did not appear until the end of the 19th century. Only by the 20th century did questions of objectivity in destination writing become standard, as trains and then planes made travel more commercial (Bertho-Lavenir, 1999, p. 59). In the past century, travel journalists have provided the essential information to help travelers make informed decisions, fulfilling a role as consumer reporters.

Within a travel context, this historical reliance on word of mouth and insider information from those who have been there may be of importance in understanding travel journalism. An underlying motivation in the tourism industry, the 'authentic experience' has, since the 1970s, been a feature of tourism discourse and, of course, marketing. Dean MacCannell launched a debate over the impossibility of this experience that infects modern society, leading travelers to look beyond the front stage into the backstage, to tinker in the local life of an otherwise exotic culture to see what *real* life is like. Ultimately, however, MacCannell deems such experiences unauthentic, because, in the end, tourism events will always be somewhat staged, even if subtly or unintentionally (1976, p. 98). Urbain explores authenticity semantically, describing the historic separation between travelers and tourists, the authenticity-seeker versus the commercial-consumer. For him, travelers walk among the natives, while tourists merely walk in the travelers' footsteps (1991, p. 103). No one *wants* to be a tourist, even if many (or all) of us are.

However impossible 'authentic' experiences might be, the idea of chumming up with the locals remains a major theme in the travel industry, with 'secret', 'hidden' and 'insider' among the words dominating museums, tours and other travel-related goods. While 'seeing the sights' like the Eiffel Tower and Notre Dame continue to be mandatory program, leisure tourists look to distinguish their travels by finding something unique and authentic – an arguably easier task for Western travelers in the depths of Cambodia than in the well-trodden Parisian landscape.

This quest for the authentic is still a major motivation for contemporary travel writers who write about their own experiences while offering often-intimate information about a destination. Travel journalism is complicated by this overlap between journaling and journalism, between the author looking into a mirror and out a window (Greenman, 2012, p. 6). Reporters in most other domains – politics, science, technology – avoid including personal anecdotes in general news stories. Even if true objectivity may be an unachievable expectation, journalists in those fields go to certain lengths to keep their personal opinions out of their stories, opting to reveal the facts and to verify rather than assert

(Kovach and Rosenstiel, 2001, p. 87). In travel journalism, subjective impressions matter. Knowing that a café exists in Paris is not enough. Readers want to know how the food tasted, how the waiters served, what the ambiance was like. Much of this reporting depends on the journalist's personality more than the café itself. What essential difference then remains between a magazine journalist who reviews a hotel and the word of mouth from a friend who stayed at the hotel? What can a journalist glean from staying in that hotel than a regular tourist – a physician, baker or lawyer cannot? Ideally, the journalist would adhere to certain professional values of transparency and accuracy that help define the profession, but that ideal is eroding, at least in travel media. The Internet is changing the game, not just in terms of how tourists organize trips, but on how they inform themselves. Why buy a newspaper for one journalist's review when the Internet abounds with sources, both seemingly professional and otherwise, that can give similar information and opinions for free?

Allowing non-journalists to pitch or even publish content is not limited to travel or even to online content. The *Zagat* guides, published since 1979, include restaurant reviews with personal comments printed alongside an official rating calculated from diners solicited for feedback. Online news sites have also flirted with audience participation on their sites either as stand-alone publications or web versions of the printed paper. This user-generated content has disoriented journalists unsure of how to cope with the invasion of amateur contributors, leading some to try to bridge the two, albeit in a limited fashion still largely controlled by editors (Singer et al., 2011). Whereas online news sites innovate their own platforms by allowing restricted audience participation through comments and forums, some travel sites have gone a step further, allowing tourists to become the sole creators of travel information in many ways. TripAdvisor, for example, would not exist without users generating the content.

The media industry is scrambling to figure out how to navigate this new territory, where pure player giants like Yelp and Wikipedia (and their 2013 travel spin-off WikiVoyage) have been growing and evolving since the early 2000s. For example, the *Fodor's* guide partnered with TripAdvisor, publishing online comments in the printed guide, removing the review from its original environment.

Travel blogs and journalism

While professionals on one end are trying to maintain a hold on travel information, with the continued printing of newspaper travel sections, magazines and travel guides, user-generated content on the other end of the spectrum has opened up publication possibilities to everyone. Online users have been given nearly free reign to publish whatever they want with minimal repercussions for false or biased information.

Somewhere between traditional, objective, professional journalism and largely reactive commentaries, between dictation and crowdsourcing, blogs have carved out a successful niche. Blogs provide quasi-journalistic information from supposedly trusted sources with honest opinions – the best of both worlds, neither mirror nor window exclusively. At the same time, there is a vetting process that ensures that bloggers maintain some sort of reputation. Unlike an individual TripAdvisor comment that does not dictate whether the TripAdvisor platform will succeed or continue to exist, blogs are more dependent on solid content that attracts and maintains an audience (Cardon and Delaunay-Teterel, 2006). Audience traffic and user comments matter to bloggers. Besides maintaining an audience, there are few constraints on bloggers' expression. By relying solely on the personal experiences of the author, and unconstrained by most external corporate or advertising limits, posts can be more intimate and personal.

My analysis of blogs, specifically blogs about Paris written in English, explores whether these writers are the unfettered 'new' journalists that many would hope. With ever-growing audiences and clout, these blogs are inching into territory previously reserved for journalists, obtaining and disseminating travel information to an active online audience.

The question whether bloggers fulfill a journalistic role has been discussed since their extensive appearance in 1999 with the introduction of free blogging software (Gant, 2007). Dan Gillmor (2006) writes about the changes during the first decade of the 21st century, as the rise of bloggers generated new forms of civic engagement. This innovative form of publication, however, often lacks the original reporting or journalistic practices that constitute professional journalism (Blood, 2003). Several writers and journalists maintain an optimistic view of blogging. Scott Gant (2007) describes bloggers as 'citizen journalists' or 'non-traditional journalists' who distribute information and engage in the same work as their professional counterparts. For Gant, the Internet erodes the artificial distinction between journalist and citizen, and communication circles back to a time when corporations didn't monopolize information dissemination. Other researchers like Le Cam (2006) suggest that bloggers are assimilating certain journalistic norms while also carving out their own niche as an alternative to the current brand name news sources.

This debate sparked many of the questions guiding my study of travel blogs: Do travel blogs and traditional media that operate in the same niche resemble each other in terms of content? Additionally, how do audiences consume these blogs and what are they looking for when reading them? Do readers make any distinction between professional sources and non-traditional forms of journalism?

Method

The study explored both the audience and the content of a sample of specific blogs about Paris, a city with no shortage of writing dedicated to it. The blogosphere in Paris is well established. Blog posts by famous local bloggers have been known to cause long lines at local eateries and bars, even if only for a few weeks. With hundreds of bloggers, tweeters, TripAdvisor commenters and so on, this particular city is an especially information-dense tourist destination. The study first analyzed the existing Paris blogosphere in English by using web-mapping tools, including a specific application called a navicrawler, to scan sites, finding the links that point both in and out of the site. The idea was to begin to understand what this particular blogosphere resembled: the number of blogs, the rate of connections and the most heavily linked sites, for example. The scan began on a well-known blog for English speakers in Paris, TheParisBlog.com, which aggregates and reproduces blog content from other, but not all, blogs. The scan allowed me to click through thousands of pages that are linked back to the original site, tagging sites that fit the profile of an active Paris-based English-speaking blog, resulting in about 160 results that can be linked back to TheParisBlog.com. Expats, mostly North American, Australian and English, are the main producers of these blogs. I tended to examine sites that only dealt with Paris over an extended period.

After the scan, a program called GEPHI was used to visualize the links, showing in a tangible manner this *'unknown territory'* of an intangible blogosphere, even if it is a simplification of the phenomenon (Plantin, 2013, p. 231). While not representative of the entire blogosphere, the analysis did reveal some of the most frequently linked blogs in the center of this blogosphere, allowing me to choose 11 specific blogs with enough content to analyze blog posts over a specific period. Several of these blogs also appeared on the Paris region's tourism website, further supporting my choice of sample sites. I chose the month of February 2012 to analyze the content of each blog.

The 95 posts during this timeframe were analyzed using a checklist of traditional journalistic practices (Kovach and Rosenstiel, 2001; Neveu, 2009; Zinser, 2001). The idea was to observe, for example, how bloggers might employ traditional journalistic practices such as lead sentences, citing official sources or attempting an objective point of view among other criteria. The checklist was divided into three main parts, dealing first with the article's content, identifying the use of first person writing, lead sentences, explanatory nut graphs and sources. It was important also to identify the category of the post, deciding whether it was a narrative or service-oriented entry, for example. In the second part of the checklist, relying heavily on Kovach and Rosenstiel's (2001) study, I looked for elements that helped situate the objective nature of the post. While it seemed excessive to expect blog posts to be objective in any concrete sense, due to their largely personal nature, I looked for the presence of facts, links to supporting websites, and a general distance between the author and the subject as indicators of an objective approach by the blogger. Additionally, I evaluated the posts for a 'practical or functional form of truth' in the article that helped define exactly what the blogger was aiming to convey (Kovach and Rosenstiel, 2001, p. 42). Finally, the third part of the checklist covered the structure of the blog, including identifying any editorial hierarchy, concrete sections or tabs, possibilities to interact with the authors, and the location of links to other sites, including social networks and advertisers. While by no means exhaustive, the list addressed some general tenets of the profession and features of institutional media that have surfaced over the last century and that were expected to appear in some, though not all of the blogs.

At the same time, to study the audience of these blogs, Paris-based bloggers helped to distribute an online questionnaire, resulting in 264 useable survey results. Using the online questionnaire platform Survey Monkey, I proposed four parts in the questionnaire with the possibility of free responses or graded responses depending on the question. The first part focused on perceptions of sources of information in general. The second part explored practices relating to researching a trip. The third part introduced the idea of authenticity in travel experiences. Finally, the fourth part requested voluntary personal information about age, gender and location of respondents. The questionnaire was posted in early February 2012 and shared on Twitter and Facebook. Several Paris-based authors of blogs I would later analyze shared the questionnaire with their readers, helping to increase the size and scope of the sample.

The respondents varied, covering a wide range of nationalities, ages, backgrounds and travel habits. A large portion consisted of armchair travelers hoping to visit Paris one day, others were study abroad students living in Paris for a few months, and still others were expats. From those who answered, I selected ten people present in Paris for further in-person interviews to explore their motivations and online practices concerning travel. Most were expatriates or were staying in Paris for more than six months, but as Urbain explains, 'An expat is just a traveler in an interrupted voyage: a person living abroad, a traveler outside of his travels' (1991, p. 273). Further studies should focus more on true temporary tourists, but the current study considered more broadly all non-French natives as tourists, as Urbain described. These interviews, along with the larger questionnaire, pointed strongly towards the conclusion that online content, especially blogs, is a primary source for secondary activities while in Paris, for all that is local or authentic.

Findings

Journalists unwanted

The analysis of the blog content demonstrated a consistent trend towards writing in which authors ignored most norms of journalism including basic form, support from official (or any) sources or an objective tone. Unsurprisingly, blogs were centered on the writer, with a heavy reliance on the first person, on personal observations, and on unsupported generalizations about Paris, Parisians or other local culture. Though two blogs did manage to write with a distanced, more objective and journalistic voice, only 27 per cent of the sample of posts included original firsthand reporting of a topic or event, as opposed to general reflections or musings. A further 31 per cent included actual supported facts, and only 20 per cent included any primary or secondary general or official source.

The blog sample also illustrated that travel blog posts in this case did not mimic many journalistic practices. In terms of content, sources – official or otherwise – were rare. Only 20 per cent cited other people, while only one post in the entire sample cited any sort of report, study or official document. Also, first person syntax ('I' or 'we') prevailed in 77 per cent of posts where it served to narrate a story or anecdote. Structurally, 54 per cent of posts did have a semblance of a lead, with only 27 per cent offering an explanatory nut graph to help situate the item in a news context. In terms of transparency, three blogs did reveal that content was either paid for or sponsored by an outside body, exhibiting some disclosure as described by Greenman (2012).

While the content produced was rarely written in a journalistic manner, the layout of the blogs did resemble the basic layout of news sites. For example, 33 per cent of posts had editorial teams behind them, supplying multiple voices and points of view like a traditional newsroom. Only one blog did not have tabs available for different sections. The rest were organized like CNN.com or BBC.com with a navigation bar supplying easy access to specific information. Like other media sites, all posts were published with links towards other websites. Finally, the blogs did offer accessible communication with the blogger. While only 7 per cent featured forums, all posts had comment spaces at the bottom with 93 per cent allowing conversation with the bloggers through Twitter, Facebook and/or other social networks.

Adhering to very few journalistic practices of verification, bloggers' trustworthiness and integrity are somewhat put into question. At first glance, journalism scholars might be tempted to hold these bloggers accountable for parading as journalists without any of the training or following any of the standard practices that help provide accurate information. The bloggers may be responding to audience expectations.

Audience expectations of authenticity

The second part of the study, questioning and interviewing the audiences of these blogs, revealed a strong tendency towards valuing, if not actively searching, the personal, amateurish touch that these bloggers employ. The respondents' main motivation for consulting blogs was seeking advice or insider knowledge from someone on the ground.

To understand audience motivations further, the concept of authenticity was interrogated in both the questionnaire and the interviews to see if it still applied to tourism in a digital world. It seems that with so much more information available than when MacCannell originally published *The Tourist*, authenticity is more important than ever. When first asked among which sources of information were the most trustworthy, respondents rated guidebooks and blogs equally, slightly above magazines, newspapers and TV shows. However, when asked later which of the five sources would provide the best information leading to the most authentic experience, blogs came out on top far above the other media. Still, the definition of what is authentic remains debatable. When asked to define authentic experiences, open responses in the questionnaire ranged from 'honest', 'non-touristy' and 'genuine' to 'French-owned', 'no other Americans' and 'hole in the wall.' This further complicates the issue of identifying an authentic experience – everyone's version of one will be different (Urry and Larsen, 2011, p. 114).

The interviewees reflected this complexity with varied responses, demonstrating a need for further study on the topic. While one woman said she finds authenticity by asking locals what to do in a town, another defined authentic experiences as those devoid of other tourists. Still another suggested that walking in the shoes of a local defines authenticity for her, even if the event revolves around something as trivial as trying to find a parking space in Paris. Authenticity was endowed to bloggers who were 'on the ground'.

The bloggers, then, were seen by respondents as foreign correspondents of sorts that distinguish themselves through prolonged experiences in the city. Audiences looked to blogs to help find those 'one of a kind' experiences or events that cannot be found in guidebooks and magazines. Most readers, nearly 66 per cent, reported reading only one to three blogs on Paris, as opposed to 10 per cent of respondents that can be regarded as blog junkies, consulting eight or more. Like news anchors or radio hosts, bloggers create relationships with their readers, so it's not surprising that there may be a certain level of loyalty among readers (Schudson, 2011, p. 118). In an open response question, nearly 80 per cent reported looking for authenticity from expats, word of mouth, Parisians and locals – including blogs written by locals or expats. The interviews further drove this point home. One interviewee discussed how she found a chocolate museum in Shanghai only because she read about it online, and another had tested a new American restaurant in Paris based on the recommendation of a favorite blogger. The unique position of an expat blogger having lived in the destination for a longer time than regular tourists or travel journalists is a significant advantage at a time when print media cannot afford to station correspondents in bureaus abroad. And readers appreciate the depth of experience and timeliness of information.

They also value the personal voice that these bloggers use. More than just a voice or writing style, the use of 'I' and personal anecdotes, a photo, a biography, shared nationality and a log of posts that help form a narrative around a writer are mentioned as positive points. The personal connection builds trust between bloggers and readers. Coupled with 'insider' content, blogs offer credible and authentic information that helps travelers be a local for a day.

Professional and paper sources not extinct

The interviews and questionnaires also demonstrated that travelers don't look exclusively online for their information. Printed travel guides with perennial information about landmarks, annual events and institutions were still considered useful. As one interviewee said, 'The Louvre is still the Louvre'. Blogs, however, were not the sites to be consulted for this type of standard information. The sample of posts in February 2012 rarely featured information about the Champs Élysées, Versailles and other sights that are important for tourists but don't figure into the daily routines of most people living in Paris.

In the questionnaire, respondents cited a plurality of sources, professional and non-professional, corporate and personal, print and online. Guidebooks, word of mouth and TV travel shows appeared alongside Google, blogs, NYTimes.com and TripAdvisor. Studies focusing on trip planning have explained different trends of when and how travelers consult certain forms of media, both journalistic and user-generated content, showing that they read user commentaries, for example, both to narrow choices before purchasing and to generate ideas before researching further details (Gretzel, 2007). There is little consensus on the specific utility of any one type of source, but practical considerations can help us begin to understand how travelers might organize their information.

Technical development and infrastructure further influenced media choice. For example, printed information is not yet outdated even though some market research notes that in 2012, 85 per cent of travelers with smartphones used them to travel for taking photos, consulting maps and finding restaurant recommendations (*Techie Traveler*, 2012). However, even major Western tourism destinations including Paris still do not have free wireless Internet everywhere, nor do they have international roaming charges diminished so much that using a cell plan is preferable to lugging around a *Lonely Planet*. Material in print, in this case, will continue to be relevant as long as digital options are not universally accessible, especially in developing countries.

A typical feature of blogs written by one person is their strong personalization and release of special 'insider' information. This positioning was another important point for reader loyalty. Some mass-appeal blogs, or even wikis, usually managed and written by several people, were more likely to provide more routine information geared to travelers looking for basic routine information. Some respondents in the interviews suggested that the diversity of voices on such a blog, however, much like a newspaper's bureau, allows for less personal connection than a solo blogger, disturbing the intimate bond to and credibility of the bloggers. While one third of the posts analyzed in this study were either written or edited by teams behind the scenes, this seemed to deter readers from returning to the blog, according to respondents. One interviewee, for example, explained that she trusts blogs with multiple contributors less then when one blogger controls the writing.

When it comes to personal blogs, however, sites that exhibit signs of commercialism, like excessive advertising or sponsored posts, are equally contentious to readers. Interviewees prefer independent-looking blogs, even though in the sample, all 11 blogs used some form of advertising on all of the posts, as well as tie-ins to other business operations (tours, apartment rentals, book sales).

Implications for bloggers and tourists

With a more personal, less journalistic approach, bloggers seem to form sustained relationships with readers, disseminating information to a motivated crowd of followers who expect personalized, timely and unique information. Readers connect to the bloggers like old friends, preparing itineraries based on posts and comments from these websites. Like other review sites, travel blogs can be understood as digital word of mouth. Through comments and social media, bloggers and readers interact, creating a conversation that both enriches readers and offers feedback to the bloggers themselves.

Beyond providing word-of-mouth information, however, these bloggers are taking up the role that was traditionally held by journalists on the ground, attending Fashion Week shows, reviewing hotels and suggesting restaurants alongside professionals. While traditional news organizations rely on branding to establish institutional news as a product that is credible and attracts audience loyalty, travel bloggers build up credibility by carefully nurturing a non-commercial identity (Ramonet, 2011). Travelers don't want to join the bandwagon and follow *Fodor's* or *Travel+Leisure*, even if these books and magazines provide solid standard information for tourist experiences. Instead, blogs more successfully cater to travelers' desire to distinguish themselves from the rest of the pack (Redfoot, 1984).

Among other sources, blogs fulfill the role to help travelers off the beaten path. Travelers trust bloggers and follow their advice. They share this information with others. As a relatively new phenomenon, even in a major city like Paris, blogs have only really blossomed over the past five to ten years. These blogger–reader relationships could have larger implications for the travel industry. Not only blog readers pay attention – tourism professionals are beginning to understand the potential behind this medium. For instance, several of the blogs studied are already on the Paris region's Office of Tourism website as points of reference for English-speaking travelers. The website, a major resource for tourists, highlights its 'favorite blogs' in English, French, Spanish and German, offering a selection of expatriate or otherwise non-professional blogs that are deemed to cater to tourists' needs. They are advertised on the site as ways to find out what's going on in Paris: 'They talk about culture, fashion, shopping, food and nightlife, providing all sorts of hot ideas and new suggestions for organizing your stays and week-ends in Paris and the region!' (Office of Tourism of the Ile de France, 2014). Increasingly, these small intimate blogs are co-opted by mainstream sites (as free content) while Internet use by travelers continues to surge.

When it comes to questions of transparency, disclosure and independence, travel bloggers might not be as trustworthy as their professional counterparts (Greenman, 2012, p. 143). In the interviews, some savvy readers explained that they were suspect when a blogger described a pricey experience, expecting a disclosure notice that the trip or hotel was paid for by someone else. As one interviewee clarified, subsidized content was not inherently a problem, but rather writers who didn't reveal such activity were untrustworthy. Only one of the 95 posts studied, however, revealed such a disclaimer. This issue ties bloggers to their traditional counterparts. Travel journalism, advertising and public relations have historically gone hand in hand, with sponsored trips and tourism-related advertising financing the bulk of such writing (see Chapter 1).

Rethinking travel information

New trends and convergence

Analyzing blogs, we can see a shift in how leisure travelers inform themselves, at least those connected to the web. Travel journalism, more clearly than other types of journalism, may actually benefit from moving online, by diversifying the topics covered, adding varying points of view, and evolving professional standards to better address audience concerns. Yet, the competition among information providers increases as well and bloggers, writers and professionals alike have to work harder to stand out from the crowd. This competition can mean less redundancy of information, as opposed to the standard news stream that often recycles similar topics and points of view across different outlets. Yet, several studies have demonstrated that online news tends to follow the same patterns of the traditional press, illustrating that a handful of topics are covered by most media outlets, while a seemingly infinite set of niche media sites cover an equally endless spectrum of niche topics (Smyrnaios et al., 2010). The logic of the 'long tail' governs the production of information as well. Central destinations in travel will probably continue to dominate the majority of sources. But with travelers looking for authentic, distinguishing experiences, bloggers are challenged to produce unique content. In the analyzed sample, there was little overlap of topics. Several blogs featured taglines like 'show you the hidden side of the City of Light' and 'your insider guide' that demonstrated a desire, even if only nominally, to differentiate themselves from the norm. Readers are constantly looking for the next new thing, and bloggers are pressed for new content. For example, when locally roasted coffee became popular in Paris, many bloggers were publishing posts in 2008 through early 2010 as the boutique coffee shops were opening. Traditional media lagged behind. *The New York Times*, for example, published an online article entitled 'In Paris Where Coffee is King', in September 2010, well after the blogosphere had begun circulating the topic.

With so many new actors, technologies and expectations, travel journalism will continue to evolve. And the shift online has rapidly forced journalists in all domains to rethink their practices. However, as Ruellan (2007) argues, a fixed definition of journalism rarely existed. He details how, over many decades in France, the US, or elsewhere, characterizations of journalism have continuingly changed.

Bloggers are creating a sustained community, as evidenced by the strong links between Paris-based blogs illustrated by my web-mapping techniques. It is becoming clear that the gatekeepers of travel information need not be professionally trained journalists. Yet overemphasizing the differences between bloggers and traditional journalists may overlook the ongoing blurring of the professional boundaries in process.

The traditional media are adapting to the web, for example, with *Fodor's* print guides partnering with TripAdvisor or *The New York Times* launching its own blogs. Simultaneously, online players are finding themselves involved in the offline world, as bloggers become contributors to traditional outlets. Freelancers, many of whom are bloggers as well, update guidebooks, write articles for *The New York Times*, or publish periodic pieces in *Condé Nast Traveler* in addition to authoring travel blogs. Many bloggers even take professionalization to a new level by publishing books, nonfiction accounts of their lives abroad, essentially memoires that collect their blogs in hardback copies (or Kindle downloads). Blogs need to be seen as a platform for travel writers to establish credibility and authenticity leveraged to a personal brand that can be extended across other media platforms in an increasingly entrepreneurial journalist environment. It remains to be seen how practices of disclosure and transparency fit into the contemporary travel journalist's profile (Ringoot and Utard, 2005, p. 46). The interviews with readers suggest that an engagement across platforms and increasing commercialization may undermine the authenticity of the amateur blogger voice the audience looked for in the first places.

Moving forward

This study demonstrated that travelers looking for an 'authentic experience' are likely to turn to blogs with a discernible voice. Bloggers, who become akin to a travel companions, are the local word-of-mouth sources that travelers can easily access. Despite lacking many standard journalistic practices in their blog posts, these authors still provide information that readers trust and accept in similar ways than the information they find in traditional media. The personal and insider nature of these platforms can provide more timely, local and backstage information. Nevertheless, institutional media, especially travel guides and magazines, persist in offering in-depth, perennial information that tourists need.

The example of blogs in Paris is just one of an endless array of studies possible on such phenomena. English-speaking expat bloggers in Paris are a niche group for a specific geographic area. It would be interesting to investigate different locations, cultures and blogger communities, such as Korean-speaking bloggers writing about Rome or French-speaking bloggers in New York.

It is difficult to conclude what the real impact of these new media will be on the tourism industry in the future. While readers in this study confirmed a preference for word-of-mouth sources in blogs, new technology like geolocation applications promise to offer alternatives to blogs and other travel media by skipping directly to locals for information. Websites like Airbnb and Vayable have entered the market, putting travelers in direct contact with locals, further eroding the need for a professional journalist to mediate travel experiences. For the moment, however, it is clear that all of these networks and websites are impacting the institution of journalism and forcing us to reposition the place of traditional travel journalism.

References

Barzilai-Nahon, Karine (2008) 'Toward a Theory of Network Gatekeeping: A Framework for Exploring Information Control', *Journal of the American Society for Information Science and Technology*, 59.9, pp. 1493–1512.

Bertho-Lavenir, Catherine (1999) *La Roue et le Stylo: Comment nous sommes devenus touristes*, Paris: Editions Odile Jacob.

Blood, Rebecca (2003) 'Weblogs and Journalism: Do They Connect?' *Nieman Report*, Fall, pp. 61–3.

Blanton, Casey (1997) *Travel Writing: The Self and the World*, New York: Simon and Schuster Press.

Bruns, Axel (2003) 'Gatewatching, not Gatekeeping: Collaborative Online News', *Media International Australia Incorporating Culture & Policy*, 107, pp. 31–44.

Cardon, Dominique and Delaunay-Teterel, Hélène (2006) 'La production de soi comme technique relationnelle', *Réseaux*, 4.138, pp. 15–71.

Duffy, Andrew and Ashley, Yang Yuhong (2012) 'Bread and Circuses: Food meets politics in the Singapore media', *Journalism Practice*, 6.1., pp. 59–74.

Gant, Scott (2007) *We're All Journalists Now*, New York: Free Press.

Gillmor, Dan (2006) *We the Media: Grassroots Journalism by the People, for the People*, Sebastopol, CA: O'Reilly.

Greenman, John F. (2012) *Introduction to Travel Journalism*, New York: Peter Lang.

Gretzel, Ulrike (2007) 'Online Travel Review Study: Role and Impact of Online Travel Reviews', *Laboratory for Intelligent Systems in Tourism*. Available at: http://www.tripadvisor.com/pdfs/OnlineTravelReviewReport.pdf.

Hanusch, Folker (2012) 'Broadening the Focus: The case for lifestyle journalism as a field of scholarly inquiry', *Journalism Practice*, 6.1, pp. 2–11.

Kovach, Bill and Rosenstiel, Tom (2001) *The Elements of Journalism*, New York: Three Rivers Press.

Le Cam, Florence (2006) 'États-unis: les weblogs d'actualité ravivent la question de l'identité journalistique', *Réseaux*, 4.138, pp. 139–58.

MacCannell, Dean (1976) *The Tourist: A New Theory of the Leisure Class*, New York: Schocken.

Marcil, Yasmine (2007) "Le lointain et l'ailleurs dans la presse périodique de la second moitié du XVIIIe siècle" *Le Temps des médias*, 1.8, pp. 21–33.

Neveu, Erik (2009) *Sociologie du Journalisme*, Paris: Editions La Découverte.

Office of Tourism of the Ile de France (2014). Favorite Blogs of Paris and its region. Available at: http://en.visitparisregion.com/favorite-blogs-290204.html.

Smyrnaios, Nikos, Marty, Emmanuel and Rebillard, Franck (2010) 'Does the "Long Tail" apply to online news? A quantitative analysis of French-speaking websites', *New Media and Society*, 12.8, pp. 1244–61.

Plantin, Jean-Christophe (2013) "D'une carte à l'autre: Le potentiel heuristique de la comparaison entre graphe du web et carte géographique", in Barats, Christine (ed), *Manuel d'analyse du web*, Paris: Armand Coin, pp.228–245.

Ramonet, Ignacio (2011) *L'explosion du journalisme*, Paris: Les Editions Galilée.

Redfoot, Donald (1984) 'Tourist Authenticity, Tourist Angst, and Modern Reality', *Qualitative Sociology*, 7.4, pp. 291–309.

Ringoot, Roselyne and Utard, Jean-Michel (2005) 'Genres journalistiques et '*dispersion*' du journalisme' in Roselyne Ringoot and Jean-Michel Utard (eds), *Le Journalisme en Invention : Nouvelles pratiques, nouveaux acteurs*. Rennes: Presse Universitaire de Rennes, pp. 21–47.

Rocamora, Agnès (2012) 'Hypertextuality and Remediation in the Fashion Media: The case of fashion blogs', *Journalism Practice*, 6.1, pp. 92–106.

Ruellan, Denis (2007) *Le Journalisme ou Le Professionnalisme du Flou*, Grenoble: Presse Universitaire de Grenoble.

Schudson, Michael (2011) *The Sociology of News*, New York: W.W. Norton and Company.

Singer, Jane; Hermida, Alfred; Domingo, David; Heinonen, Ari; Paulussen, Steve; Quandt, Thorsten; Reich, Zvi; Vujnovic, Marina (2011) *Participatory Journalism: Guarding Open Gates at Online Newspapers*, West Sussex: Wiley-Blackwell.

Techie Traveler (2012, March 23) Available at: http://blog.lab42.com/techie-traveler.

Urbain, Jean-Didier (1991) *L'idiot du voyage: Histoires de touristes*, Paris: Petite Bibliothèque Payot.

Urry, John and Larsen, Jonas (2011) *The Tourist Gaze 3.0*, London: Sage.

WWWmetrics (2014) *Growth of the Travel Industry Online*. Available at: http://www.wwwmetrics.com/travel.htm

Zinser, William (2001) *On Writing Well: The Classic Guide to Writing Nonfiction*, New York: Harper Collins.

6

First-Person Singular: Teaching Travel Journalism in the Age of TripAdvisor

Andrew Duffy

Introduction

The Internet is a foreign country: they do things differently there. So in a digitized world the first source of information for travelers is often online – blogs and user-generated content with personal reports on holidays and journeys, destinations and attractions, hotels and restaurants (Casaló, Flavián and Guinalíu, 2011). The information they find there is rarely written by experts; instead it is the voice of experience. As a result, travel journalism – for both writers and readers – is influenced by first-person writing in the form of blogs, social media and traveler information-sharing platforms such as TripAdvisor and Lonely Planet's Thorn Tree pages. This is likely to impact the travel journalists of the future, who have not come of age backpacking with a well-thumbed copy of the Rough Guide, but journey instead with the opinions of millions on the Internet at their fingertips (Hofstaetter and Egger, 2009). This situation has already altered the marketplace. Anecdotally at least, newspapers want first-person travel journalism that replicates the experiential authenticity of its online relatives.

This chapter explains first how experience has replaced expertise as the common currency of travel writing. This leads to two questions: how best to teach journalism students to operate within a new paradigm dominated by this first-person style of writing, and what are the implications if students' travel-writing benchmarks are blogs and review sites? Second, the chapter proposes that when online travel writing celebrates the amateur's experience over the professional's expertise, there is a need to educate travel journalism students so that their own travel experiences go beyond those of the amateur, and so that their reporting transcends mere self-concern to engage on equal terms with the host nation and the assumptions of their readers.

Challenges for (travel) journalism education

The reason for scholarly attention to travel journalism is well established; it gives an insight into how foreigners are represented for a home audience, which can be a bellwether for how minorities are represented in the media (Cocking, 2009; Fürsich and Kavoori, 2001; Hanusch, 2010). Good (2013, p. 296) states that 'travel journalism – just like "serious" forms of journalism – warrants attention as documentation of the shared assumptions between journalists and readers about what representations are relevant from beyond their borders'. Extending this idea to the classroom raises the question of how students can be educated to consider the ideologies and ethics of travel writing, which includes an awareness of their own perspective and biases, and knowledge of others' perspectives.

Demand for such journalism education is not new. Deuze, for example, suggests a globally aware journalism course which would build sensitivity to international issues, and which could be a solid starting point for a course in travel writing. In such a course, he explains,

> students would be confronted in all matters by the cross-cultural or transnational nature of what they are learning ... Next to globalization, one could also think of the multicultural society, featuring themes such as social and cultural complexity, inclusivity and diversity awareness. (Deuze, 2006, p. 27)

Yet, as Holm (2002, p. 69) has observed, journalism courses that 'reflect on national stereotypes, global affairs and global-social transformations are few and far between.' Adding urgency to the need for such courses, competition from amateur online travel information now obliges professional travel journalists to differentiate themselves if they expect to continue being paid for what millions of netizens post for free.

Travel blogs and 'ours'

Today, travelers consult the Internet for information, looking at travel blogs and user-generated content sites such as TripAdvisor and Agoda. As the content on the latter is created by us, the general public, this chapter uses the acronym of OURS, which stands for Online User

Review Sites. Reading travel blogs and reviews is a popular form of holiday planning (Gretzel, Yoo and Purifoy, 2007), and in 2005 the Pew Internet and American Life Survey reported that searching for travel information is one of the most popular online activities (O'Connor, 2010). The Internet's value as a travel planning tool has grown, and 66 per cent of US travelers said they used it in 2009, up from 35 per cent in 2000 (Li and Wang, 2011).

Blogs, or online diaries, first appeared in the mid-1990s, and travel blogs were early incarnations. While defining a blog is a vexed undertaking (Garden, 2012), this chapter uses a definition that is both structural and functional, but loose, in that the genre of blog under discussion is a frequently updated online personal report on travel, presented in reverse chronological order, with the aim of sharing travel experiences. (Parenthetically, blogs, as a shortening of 'web log', have a pleasing etymological connection with travel. A log or logbook was a record of a voyage based on the captain throwing a log of wood tied to a knotted rope over the stern of a ship, to gauge speed through the water.) Blogs have been celebrated for their authenticity, honesty and subjectivity, posting personal experiences (Schmallegger and Carson, 2008); they are criticized when they are 'incoherent, unstructured, random ramblings' and of 'questionable value' (Akehurst, 2009). They can be solitary efforts, or they can involve many writers on sites such as travelblog.org, travelpost.com or travelpod.com. They are commonly read by young Internet veterans, the people who form the next generation of travel journalists.

Bloggers share experiences more than emotions (Volo, 2012), and the most common topics for travel bloggers are weather, food, transport and regional stereotypes (Schmallegger and Carson, 2008). Tse and Zhang (2013) found that they covered attractions (40 per cent), food (20 per cent), shopping (11 per cent), transport (10 per cent), people (9 per cent) and accommodation (8 per cent). What is less clear is how much blogs directly affect travel; while Hofstaetter and Egger (2009) found that backpackers depend on ad hoc information and authentic reports to decide where to go, Volo's (2010) research did not show blogs influencing travel plans.

Blogs exist in an area that has long been the domain of travel journalism: that of infotainment. Deuze (2001) notes that the culture and entertainment industries converge, challenging journalistic ideologies of objectivity. While infotainment is not respected among news reporters, it is encouraged in travel journalism, which readers expect to be informative and entertaining, and which is often written from a subjective first-person viewpoint. This places it close to the realm of the blogger, as most blogs alternate between first person and second person, both drawing the reader in and co-creating an identity for the writer (Bosangit, McCabe and Hibbert, 2009; Hofstaetter and Egger, 2009). Travel blogs usually map a personal journey (Azariah, 2011), and are concerned with a personal experience written in the first person ('here is what I did') rather than the more generalized third-person form often found in travel journalism ('here is what there is'); although both contain an implicit second-person appeal to a reader ('here is what *you* can do'). The choice of first, second, or third person has significance beyond aesthetics or being un/sympathetic towards a foreign destination, and one study found that the use of the first person in travel journalism by university students is associated with being critical or negative about the host nation (Duffy, 2012).

If blogs are personal records, OURS represent a more collective endeavor. Travel recommendation sites such as TripAdvisor are attractive for travelers wishing to compare notes on their experiences. The site was a reaction to the corporate scramble for the Internet in the late 1990s that offered a more efficient means of organizing and offering tourism and travel products; airlines and car hire companies built e-commerce systems to connect customers directly to their reservation systems, while new tourism eMediaries such as Expedia.com grew as forms of web-based travel agents (Buhalis and Licata, 2002). At the same time, the rise of social media was turning the Internet into a place for personal communication. The demand for unbiased, non-commercial information was met by TripAdvisor when it launched in 2000. Over the years it has incorporated more reporting features alongside the reviews, such as photographs and videos, and social features, such as the Traveler Network in 2007 (Miguéns, Baggio and Costa, 2008) and a link to Facebook in 2012. It does not offer a reservation facility, however, and newer arrivals to the scene include Booking.com, Travelocity and Agoda.com, which allow travelers to both book hotels and subsequently write a review.

The site's strength lies in the volume of information it holds; TripAdvisor (2013) has 100 million reviews about 2.7 million hotels, attractions and destinations. While individual reviews for a specific hotel or attraction are likely to differ from each other, the user will read several and decide on balance what the hotel or attraction is really like. OURS' weakness lies in the uncertain provenance of the reviews, and despite using software to identify and remove fraudulent postings, TripAdvisor has come under attack for featuring fake reviews that misrepresent hotels (O'Connor, 2010; Smith, 2011). Reviewers are motivated to contribute to OURS by either a combination of altruism and self-interest, or status seeking to appear to be sophisticated, well-traveled experts (Lampel and Bhalla, 2007), or even to reward a service provider and to help other travelers (Yoo and Gretzel, 2008). OURS help deal with the uncertainty of buying an experience good such as food, wine or travel which cannot be tested in the way that sunglasses or a camera can (de Vries and Pruyn, 2007; Mudambi and Schuff, 2010). As a result, they play a greater part in traveler decision-making compared with other information sources, including travel journalism (Casaló, Flavián and Guinalíu, 2011).

New journalists and journalism education

While OURS and blogs are a direct challenge for commercial information sources, they also threaten travel journalists who must find ways to differentiate their work from this tsunami of information if they wish to be paid for content that others are posting for free. This situation calls on travel-writing teachers and journalism educators to help students prepare to work in an environment transformed by these newcomers. One starting point might be Deuze's (2001) guidelines for journalism education, which identified core characteristics that include the following: journalists provide a public service; they are credible because they are neutral, objective and fair; they can enjoy editorial autonomy and freedom; they are concerned with facts and validity; and they have a sense of the ethics and legitimacy the work entails. High ideals, but given the international, cultural and social sensitivities often required of travel journalism, they are a solid foundation; and it is not impossible to square these ideals with the expectations of first-person, experience-based travel writing.

Moreover, adding to the challenge to travel journalism, blogs and OURS equally exhibit some of these characteristics: they provide a public good; if individually they are not credible, neutral or fair, taken en masse they allow a reader to form a conclusion based not so much on objectivity as on collective subjectivity; they enjoy editorial autonomy and freedom, often greater than that of advertising-driven publications; they are concerned with opinions rather than facts, but experiences are often presented as fact; and their ethical stance often understands sharing as a service to the common good, although this is tempered by sometimes less altruistic motivations for posting.

The first step in building a travel journalism course that takes these issues into consideration is to acknowledge that writing based on personal experience demands an understanding of the person to whom those experiences happened and of how their identity might have colored that experience. In demographic terms, journalists tend to come from the educated middle classes, and concern has been voiced that college-educated journalists may lose the common touch and have problems reporting on the 'ordinary people' (Frith and Meech, 2007). If travel journalists are primarily taken from or inducted into the middle classes through education, it is incumbent on that education to consider what implications this might have for the way they write about foreign cultures, as social class carries assumptions of status. For the same reason, students might be encouraged to reflect on the expectations of readers who are also likely to be from the educated middle classes. Education clearly matters: Weaver (1998, p. 478) set out to identify globally consistent attitudes, norms and beliefs among journalists, but concluded that rather than global journalistic principles, journalists' backgrounds, 'including their educational experiences' had some relationship to what is reported. In addition, with growing professionalization of the industry, journalism schools and university courses have flourished. Based on a survey of journalists in 28 countries, an average of 82 per cent of them have a college degree, although there is a marked variance across countries. However, this figure halves for specifically journalism graduates, to 42.5 per cent (Willnat, Weaver and Choi, 2013).

To consider these issues, this chapter takes Fürsich and Kavoori's (2001) three perspectives for the analysis of travel journalism: periodization, power and phenomenology. Their article was written before the advent of online peer-to-peer travel information sharing, however, and they could not have foreseen the effect of websites such as TripAdvisor and the rise of the common traveler as a voice of wisdom. Thus, this chapter takes their framework and considers first how blogs and OURS might affect what Fürsich and Kavoori observed, and second the implications for the education of travel journalists.

Issues of periodization: learning to negotiate the new landscape

Fürsich and Kavoori (2001) explain tourism as an icon of modernity, defined and created by four developments: the rise of the leisure society; authenticity as a marker of a truth and purity found outside industrialized, urban societies; the division of the social environment into separate units of experience; and the impact of mediating technology that selects and presents images and anecdotes that define a place and a people. Moreover, they consider nationalism as a marker of modernism as nation-states seek to create themselves as modern countries by defining their identities as tourism sites, and the resulting impact on these national identities created by decisions of tourism boards. In addition, they also discuss a postmodern idea of tourism, whereby tourist culture is a hybrid of the traditional ways of a culture and the new touristic, voyeuristic demands of another to create ritual performances for the 'tourist gaze' (Urry, 1990).

Travel blogs and OURS affect each of these observations, and in turn offer new avenues for teachers of travel journalism. First, the rise of the leisure society has found a new voice as millions of leisure travelers share experiences. As their opinions are sought before, during and after a journey, they become a normative part of travel. Travel journalists can use them to find which attractions are popular, to gauge travelers' attitudes towards a place and to find more information after a trip. Students must learn the ethics and etiquette of using this new source, questioning the transferability of personal experience, the acceptability of taking information from such a potentially unreliable source, and whether it can guide choice of subject and frame.

Second, travel blogs and OURS present themselves as authentic experience, an antidote to (untrustworthy) commercialized travel information. They may lack the professional credibility of journalism, but they make up for it in volume and the wisdom of crowds (Surowiecki, 2005). This challenge raises the game for travel journalists, who must develop the literacy required to negotiate this new landscape, which involves an appreciation of four forms of authenticity: that expected by the reader; that promoted by the tourist board; that experienced by the journalist; and that lived by the inhabitants of the host nation. The challenge for travel journalists is to blend those four forms of authenticity as a counterweight to simple first-person narrative of experience.

This is not a new skill – travel journalists have long had to balance authenticities of how the tourist office presents a destination with the reality that they themselves encounter. This balancing act is in turn bound up with authenticity and nationalism as a marker of modernism. For example, an 'authentic' travel experience can conform more to the tourism board's projected image than to life lived by the locals. So a trip to Australia is not complete without kangaroos and a 'barbie', even though most Australians' lived experience is more likely to involve such mundane things as traffic lights and TV remotes than koalas and boomerangs. An awareness of these pressures could be instilled in travel journalism students to make them notice if they are swayed by someone else's commercial imperative. For example, the use of the second-person 'you can/should/must see/do/buy this while you are in Montevideo/ Madrid/Macau,' can be either the generous tone of friend advising friend, or the self-serving pitch of a salesperson directing customers towards the till.

Authenticity also relates to the impact of technologies such as blogs and OURS, which create or influence writers' and readers' perception of travel. User-generated content, for example, reveals and defines the tourist gaze. For example, a recent study of reader-submitted photographs on the *New York Times'* website compared two tropes: 'enduring' travel representations – which replicate traditional, romantic, aesthetic tourist experience – and 'emergent' ones – a more playful, postmodern, or self-referential take on tourism Good (2013). Teaching a writer to identify such tropes would be a step towards looking at the host nation with fresh eyes. This implies risk, as all new practice does; there is safety in stereotypes, and young writers may feel comfortable using them or believe that certain ways of writing 'sound like' travel journalism. Ideally, a course would raise this, and give them the choice to challenge conventions or clichés that represent a lack of thought, or to use them to forge connections with readers reassured by such comfort food.

Finally, blogs and OURS can be seen as a form of postmodern journalism in which ordinary people gather information and make up their own minds rather than being directed towards a fixed truth. Bloggers see truth as 'emerging from shared, collective knowledge ... created collectively rather than hierarchically' (Singer, 2007, p. 85). The collective nature of OURS and, to a lesser extent, blogs, highlights this phenomenon. At the same time, travel journalism is a postmodernist hybrid of authenticities: that demanded by the organizer; that expected by the reader; that experienced by the writer; and that lived by the people of the host nation. Without directing a writer to favor any one of these, a travel journalism course might show how a focus on each has a different effect and suggest circumstances in which each might dominate.

Indeed, contemporary tourism epitomizes many aspects of postmodernity, where boundaries between real and staged blur, where rituals become performances to be observed rather than lives to be lived (MacCannell, 1973). Again, this raises the question of what is authentic for a generation of travel writers raised on blogs and OURS, as issues of truth for blog readers are subsidiary to respect for experience, opinion and authenticity. Facts, truth and the reality of journalism are called into question by modern ideas of relativity and subjectivity (Zelizer,

2004). If truth is relative, and the objective norms of journalism are challenged by blogs and OURS, then there is comfort in reporting on personal experience that writers know to be true, so that authenticity takes the place of empirical, confirmed 'truth'. Experience replaces expertise. What, then, is left to teach travel journalists whose professional credibility was previously based on expertise as a writer, if not as a traveler? One answer might be to teach students how to report on personal experience but with a view to others – the reader on behalf of whom they travel, and the host nation to whom they are indebted for hospitality and the story itself – in order to give the reported experience a depth that takes it beyond solipsism.

Issues of power and identity: learning from BRICS and beyond

Since the 1970s and Said's *Orientalism*, concerns over post-imperialism and postcolonialism have run deep through academic thought on travel writing. As Fürsich and Kavoori (2001) note, postcolonial scholars argue that media discourse including travel writing (and journalism) subjugates others discursively (just as imperialism controlled the 'native' by colonization) in a context in which tourism controls economically and reinforces existing inequalities. For them, travel journalism is seen as a form of cultural inscription rather than cultural description.

This postcolonial academic discourse is based on tourism being 'the West visiting the rest' and maybe a reinvigoration of fading power relations inherent in that. Increasingly though, as Fürsich and Kavoori recognize, tourists travel from East to West and East within East, while the rising wealth and cultural power of Asian tiger economies and the BRICS group is coupled with downturn in Europe and the US. The next generation of journalists must re-evaluate relationships of power implicit in travel writing that is now situated in other geopolitical contexts and impacted by the increasing diffusion of Internet technologies. In addition, it is difficult to imagine a type of travel journalism that goes beyond 'othering' since encountering an 'other' is its main justification for existence (Hanusch, 2010) – a point that was arguably missed by some cultural studies critiques on travel journalism. Travelers – and readers of travel writing – want to see the 'exotic' and find something new; in an earlier analysis of student travel journalism I found that strangeness was written about more often than familiarity (Duffy, 2012). Students have to understand that writing about the 'other' per se is not of concern. Rather, the problem arises when 'otherness' is represented as less sophisticated, less developed, less civilized and less valuable.

One approach I have used is to show students how travel guides write about their home nation, so that they can reflect on their own discomfort when they see their country reported; this alerts them to a need to be aware of any judgmental stereotypes they themselves may have. Beyond reflecting on their own writing practices, students could probe the effect of their writing on their readers. Travel writers are part of an environment in which their readers are also creators. Just as students need to contemplate their own assumptions, they might consider how their writing could reinforce the assumptions of their audience. This can lead in general to a more critical reflection on the 'realities' media create. As Curran has suggested:

> The media ... provide explicit frameworks of explanation, as well as tacit understandings based on associations of ideas, evocative images, 'natural' chains of thought ... The principal way in which the media can influence the public is not through campaigning and overt persuasion but through routine representations of reality. (Curran, 2002, p. 163)

Giving readers what they want may be viewed as a skill or as selling out: Holland and Huggan (2000, p. viii) conclude the latter and argue that successful travel writers identify what their readers want and 'pander to their whims'. They criticize these writers for holding 'complacent, even nostalgically retrograde, middle-class values' (p. viii). And yet, journalism is done for an audience and must connect with that audience's worldview. To be relevant to a reader, a Marrakech souk must be described either in terms of difference from the mall that is the reader's usual retail experience, or in terms of similarity. Once again, this is not unique to journalists; bloggers, too, construct personalized meanings of their experiences based on their own cultural background, often drawing comparisons between home and away and using these comparisons to build self-understanding (Bosangit, McCabe and Hibbert, 2009).

At the same time, however, the international scope of blogs and OURS offers a more globalized perspective and hence can move journalism students away from writing for a specific national audience, instead considering issues of identity raised by the globalization of news outlets (Holm, 2002) and the Internet. When a Canadian travel journalist writes in the *Toronto Star*, it is for a local audience; but when a Canadian traveler posts a review on TripAdvisor, it is for a global audience. The posts may still resonate more with readers from a similar culture, but the fact is that the Internet has blurred the role and identity of the audience. In addition, travel information now comes from writers of various nationalities and meets online, challenging the idea of a local audience and a local worldview, as audiences are exposed to opinions and viewpoints from beyond their community.

This globalization-by-Internet triggers a more fundamental discussion on the relationship between journalist and audience usefully initiated by journalism teachers. Students could deliberate on whether travel writers' primary role is to fulfill the demands and attitudes of the community for which they write, or to pursue a more normative role of changing attitudes following a socially conscious approach, or to take a third way, reacting sympathetically to the voice of their community while simultaneously educating readers to be sympathetic to the countries they visit. To achieve this, teachers must help students interrogate their own assumptions and viewpoints, those of the readers, and those of the host nation.

One strategy is to draw attention to the risk of 'othering' in travel journalism, and to give space for counteraction (Fürsich, 2002). Duffield (2008, p. 108) argues that 'cultural awareness and cultural learning have come to be recognized as centrally important in the educational process,' and that is particularly relevant to travel journalism. A subsequent question is whether it is even possible to represent one culture to another fairly; but reflective awareness is a step in the right direction. Faced with the dilemma inherent in describing the other for a paying public, 'only self-reflective and critical approaches towards traditional ritualistic reporting and production strategies can help to disentangle problematic media representations' (Fürsich, 2002, p. 57). The questions Fürsich and Kavoori (2001, pp. 161–2) ask are still valid and should be included in a course on travel journalism: What are the dominant modes of representation of the country visited? Who benefits from the framing of cultural encounters? And how will that change as new tourism constituencies construct the travel experience?

'Othering' implies relations of power: online travel writing represents shifts in two forms of this power and both echo across journalism education. First, travel blogs and OURS can come from East or West, global North or global South, and reflect economic rather than cultural power. Anyone who travels and has an Internet connection can contribute. Journalism and journalism education, likewise, are not the prerogative of the West, and emerging economic powers can choose their own approach. On one hand, just as the Western countries which led the post-imperial tourism boom of the 1960s sought to define (and reassure) themselves through often less-than-flattering depictions of other nations, young travel journalists from newly confident emerging economies may follow a similar path. History need not repeat itself, however, and courses in travel journalism in emerging countries can equally set their own agenda following this power shift. Journalism students in former second-world countries of Eastern Europe with roots in state-controlled Communist propaganda (Bao, 2005) might confront how historical power inequalities resulting from their own histories affect travel-writing attitudes towards both first-world and developing nations. Students in emerging, former third-world countries can question whether to portray others that they encounter as lesser in order to demonstrate their own nascent power and national identity; or to learn from their own postcolonial experience and produce reflective travel journalism that contextualizes tourism within the history of the host nation. These fledgling approaches are alternatives to the economically friendly but uncritical standard route in courses that follow commercial rather than cultural imperatives.

Second, blogs and OURS have changed the power dynamics within the travel and the media industries. Information that was once controlled by commercial interests and government agencies now has a platform on which it is liberated from these controls. Writing that was once the preserve of the journalist – although increasingly done by celebrities or family and friends of the travel editor (Moss, 2008) – is now delivered by an army of everyone. The power has moved from professional to amateur, from the expert to the experienced. For journalism education, this raises questions of the intrinsic value of travel journalism. If its central worth has been to offer an objective (or at least a credibly subjective) report on experience, that contribution has now been superseded by blogs and OURS. If its value lay in counteracting the effects of commercial information (and this is highly contested) then that, too, is done in greater volume by OURS, using first-person opinions. Hence, a course might usefully guide students towards differentiating their writing from online sources, by moving beyond simple first-person reports on experience, to create something that clearly indicates expertise *combined* with experience. Students could adopt a more factual, third-person form of writing, more akin to ethnography than a holiday journal, and adopt a more sympathetic, inclusive form of writing with more local voices that does service to the host nation, the reader and the journalism profession. Alternatively, they might move into a more literary niche, where quality of writing, emotional heft and narrative power place it on a different level from blogs and OURS.

Issues of phenomenology and experience: learning about authenticity

Not all travelers are the same, and Fürsich and Kavoori (2001) also consider the range of experiences of visitors interacting with a host nation. Tourists can want cultural or recreational travel, they can be organized or drifters, and their level of cross-cultural interaction depends on what kind of experience they seek and achieve. Beyond this there are post-tourists and un-tourists, who travel but reject standard tourist behavior in favor of environmentally friendly, presumably more sophisticated, forms of travel. They may understand that they cannot have an authentic experience, and instead escape tourism as 'staged authenticity' (MacCannell, 1973) by seeking their type of authenticity in cultural back regions. They distance themselves from regular holidaymakers who dwell in the front regions of presented tourist experiences.

In another way, this front/back split is the realm of blogs and OURS, and should be considered by students wishing to progress beyond the 'what I did on my holidays' school of travel writing. By their very nature, blogs and OURS elevate the personal, authentic experience with its pitfalls of self-indulgence and irrelevance. At the same time, they often counter the commercial, the inauthentic, the front, and they strive to offer the back-room view (in Goffman's terms). Yet, while they purport to be back-room reviews of certain sites or specific hotels, they are simultaneously front-end performances by writers taking on a persona to express their opinions. A travel journalism course can use this idea to raise questions about the subjective nature of experienced tourism authenticity in general, including representations of non-commercial authenticity off the beaten track.

Education as exploration

Protected as it is from the commercial rigors of travel journalism, education can engage in a discussion of values and practices that the profession cannot. In a time of upheaval in the industry, a course on travel journalism can function as a laboratory for new journalistic approaches rather than unthinkingly replicating outdated practices.

The goal of such reflection is to offer alternatives and foster innovation. This approach presents a counterweight to the tendency in some academic analysis to focus on the problems of travel journalism instead of its potential. This dichotomy is similar to value-laden typologies of experience which elevates one over another, with 'back' being better than 'front', 'authentic' better than 'popular', 'traveler' better than 'tourist', or even 'journalist' better than 'blogger'. The preference tends to be to attain a higher plane of experience, of being, traveling, writing or thinking. Popp speaks of the 'boundary-crossing possibilities of travel' (2010, p. 133) and this accounts for both its transformative power when a boundary *is* crossed and for the ire of the academy when instead it reinforces existing social, cultural, gender and political norms. As Holland and Huggan (2000, p. 4) explain, 'travel writing can arguably be seen ... as having transgressive potential: in allowing the writer to flout conventions that exist within his/her society, it subjects those conventions – those often rigid codes of behaviour – to close critical scrutiny'.

The potential to transgress is not the same as the requirement to transgress, however, and this should also be confronted by travel writing students when considering such typologies of experience. Travel journalism can be transgressive, but it can also reassure. It is liminal, occupying a boundary between home and away, creating an opportunity to see and be an 'other' while remaining rooted in the status quo. It reassures and challenges at the same time. Travel journalism that reassures – that is, keeps home norms as the primary reference point – attracts more criticism than travel journalism that challenges – that is, keeps the new and the foreign as the primary reference point. Yet both approaches are valid; a course that challenges assumptions without establishing new rigid normative frameworks will allow space for the next generation to find its own voice.

To be a travel writer is to travel a second time. Just as a traveler can choose adventure or safety, challenge or comfort, strangeness or familiarity, so the journalist can choose between risk and cliché, words that question and words that reassure. To raise awareness of these issues in the minds of a new generation of journalists who have grown up with an online world of opinions and information from travelers from all nations *to* all nations, is to bring hope that travel journalism can be a passport to new ways of seeing and of reporting and practicing journalism: an approach that offers choice rather than preferring one over the other. Blogs and OURS might be explored as turning experience into a new form of expertise. Travel journalism education could do well to encourage reflection that allows students to move beyond reports on individual experience. Their expected new expertise lies in seeing not just with self-regarding eyes, but through the eyes of the reader, the commercial (media and tourism) providers, *and* the inhabitants of the host nation while balancing those differing viewpoints. It is a high ideal, but a skill worth mastering to gain access to this often sought-after, glamorous and exciting genre in journalism.

References

Akehurst, Gary (2009) 'User Generated Content: The Use of Blogs for Tourism Organisation and Tourism Consumers', *Service Business*, 3.1, pp. 51–61.

Azariah, Deepti Ruth (2011) 'Whose Blog is it Anyway? Seeking the Author in Formal Features of Travel Blogs'. Eleventh Humanities Graduate Research Conference, 10 November 2010, Perth, Australia. Available at http://espace.library.curtin.edu.au/R?func=dbin-jump-full&local_base=gen01-era02&object_id=171428

Bao, Jiannu (2005) *Going with the Flow: Chinese Travel Journalism in Change*. Unpublished PhD Thesis. Brisbane, Queensland University of Technology.

Bosangit, Carmela, McCabe, Scott and Hibbert, Sally (2009) 'What Is Told in Travel Blogs? Exploring Travel Blogs for Consumer Narrative Analysis', in Wolfram Höpken, Ulrike Gretzel and Rob Law (eds) *Information and Communication Technologies in Tourism 2009*. Springer: Vienna, pp. 61–71.

Buhalis, Dimitrios and Licata, Maria Cristina (2002) 'The Future of Tourism Intermediaries', *Tourism Management*, 23, pp. 207–20.

Casaló, Luis, Flavián, Carlos and Guinalíu, Miguel (2011) 'Understanding the Intention to Follow the Advice Obtained in an Online Travel Community', *Computers in Human Behaviour*, 27, pp. 622–33.

Cocking, Ben (2009) 'Travel Journalism: Europe Imagining the Middle East', *Journalism Studies*, 10.1, pp. 54–68.

Curran, James (2002) *Media and Power*, London: Routledge.

Deuze, Mark (2001) 'Educating "New" Journalists: Challenges to the Curriculum', *Journalism & Mass Communication Educator*, 56.4, pp. 4–17.

Deuze, Mark (2006) 'Global Journalism Education: A Conceptual Approach', *Journalism Studies*, 7.1, pp. 19–34.

de Vries, Peter and Pruyn, Ad (2007) 'Source Salience and the Persuasiveness of Peer Recommendations: The Mediating Role of Social Trust', in Yvonne de Kort, Wijnand Ijsselsteijn, Cees Midden, Berry Eggen and B.J. Fogg (eds), *Persuasive Technology*. Stanford, CA: Springer, pp. 148–59.

Duffield, Lee R. (2008) 'Student Reporting Abroad: An International Programme Called Journalism Reporting Field Trips', *Pacific Journalism Review*, 14.2, pp. 102–22.

Duffy, Andrew (2012) 'Out of their Comfort Zone: Student Reactions to Cultural Challenges While Travel Writing', *Asia Pacific Media Educator*, 22.1, pp. 1-13.

Frith, Simon and Meech, Peter (2007) 'Becoming a Journalist: Journalism Education and Journalism Culture', *Journalism*, 8.2, pp. 137–64.

Fürsich, Elfriede (2002) 'How Can Global Journalists Represent the "Other"? A Critical Assessment of the Cultural Studies Concept for Media Practice', *Journalism*, 3.1, pp. 57–84.

Fürsich, Elfriede and Kavoori, Anandam (2001) 'Mapping a Critical Framework for the Study of Travel Journalism', *International Journal of Cultural Studies*, 4.2, pp. 149–71.

Garden, Mary (2012) 'Defining Blog: A Fool's Errand or a Necessary Undertaking', *Journalism*, 13.4, pp. 483–499.

Good, Katie Day (2013) 'Why We Travel: Picturing Mobility in User-Generated Travel Journalism', *Media, Culture & Society*, 35.3, pp. 295–313.

Gretzel, Ulrike, Yoo, Kyung Hyan and Purifoy, Melanie (2007) 'Online Travel Review Study: Role and Impact of Online Travel Reviews', Laboratory for Intelligent Systems in Tourism. Available at http://www.tripadvisor.com/pdfs/OnlineTravelReviewReport.pdf

Hanusch, Folker (2010) 'The Dimensions of Travel Journalism', *Journalism Studies*, 11.1, pp. 68–82.

Hofstaetter, Christof and Egger, Roman (2009) 'The Importance and Use of Weblogs for Backpackers', in Wolfram Höpken, Ulrike Gretzel and Rob Law (eds) *Information and Communication Technologies in Tourism 2009*. Springer: Vienna, pp. 99–110.

Holland, Patrick and Huggan, Graham (2000) *Tourists with Typewriters: Critical Reflections on Contemporary Travel Writing*, Ann Arbor: University of Michigan Press.

Holm, Hans-Henrik (2002) 'The Forgotten Globalization of Journalism Education', *Journalism & Mass Communication Educator*, 56.4, pp. 67–71.

Lampel, Joseph and Bhalla, Ajay (2007) 'The Role of Status Seeking in Online Communities: Giving the Gift of Experience', *Journal of Computer-Mediated Communication*, 12.2, pp. 434–55.

Li, Xu and Wang, Youcheng (2011) 'China in the Eyes of Western Travelers as Represented in Travel Blogs', *Journal of Travel and Tourism Marketing*, 28.7, pp. 689–719.

MacCannell, Dean (1973) 'Staged Authenticity: Arrangements of Social Space in Tourist Settings', *American Journal of Sociology*, 79.3, pp. 589–603.

Miguéns, Joana, Baggio, Rodolfo and Costa, Carlos (2008) 'Social Media and Tourism Destinations: TripAdvisor Case Study', paper presented at IASK ATR2008 (Advances in Tourism Research 2008), Aveiro, Portugal, 26–28 May 2008. Available at: http://www.iby.it/turismo/papers/baggio-aveiro2.pdf

Moss, Chris (2008) 'Travel Journalism: The Road to Nowhere', *British Journalism Review*, 19.1, pp. 33–40.

Mudambi, Susan and Schuff, David (2010) 'What Makes a Helpful Online Review? A Study of Customer Reviews on Amazon.com', *MIS Quarterly*, 34.1, pp. 185–200.

O'Connor, Peter (2010) 'Managing a Hotel's Image on TripAdvisor', *Journal of Hospitality Marketing & Management*, 19.7, pp. 754–72.

Popp, Richard K. (2010) 'Domesticating Vacations: Gender, Travel and Consumption in Post-war Magazines', *Journalism History*, 36.3, pp. 126–37.

Schmallegger, Doris and Carson, Dean (2008) 'Blogs in Tourism: Changing Approaches to Information Exchange', *Journal of Vacation Marketing*, 14.2, pp. 99–110.

Singer, Jane B. (2007) 'Contested Autonomy: Professional and Popular Claims on Journalistic Norms', *Journalism Studies*, 8.1, pp. 79–95.

Smith, Oliver. (2011, September 13) 'TripAdvisor Removes "Reviews you can Trust" Slogan from its Website', *The Daily Telegraph*. Available at http://www.telegraph.co.uk/travel/travelnews/8760616/TripAdvisor-removes-reviews-you-can-trust-slogan-from-website.html

Surowiecki, James (2005) *The Wisdom of Crowds*, New York: Random House.

TripAdvisor (2013) About us. Accessed at http://www.tripadvisor.com/PressCenter-c6-About_Us.html on 25 September 2013.

Tse, Tony and Zhang, Elaine Yulan (2013) 'Analysis of Blogs and Microblogs: A Case Study of Chinese Bloggers Sharing their Hong Kong Travel Experiences', *Asia Pacific Journal of Tourism Research*, 18.4, pp. 314–29.

Urry, John (1990) *The Tourist Gaze*, London: Sage.

Volo, Serena (2010) 'Bloggers' Reported Tourist Experiences: Their Utility as a Tourism Data Source and their Effect on Prospective Tourists', *Journal of Vacation Marketing*, 16.4, pp. 297–311.

Volo, Serena (2012) 'Blogs: "Reinventing" Tourism Communication', in Marianna Sigala, Evangelos Christou and Ulrike Gretzel (eds), *Social Media in Travel, Tourism and Hospitality: Theory, Practice and Cases*. Farnham, UK: Ashgate, pp. 149–163.

Willnat, Lars, Weaver, David H. and Choi, Jihyang (2013) 'The Global Journalist in the Twenty-First Century', *Journalism Practice*, 7.2, pp. 163–83.

Weaver, David H (ed.) (1998) *The Global Journalist*, New Jersey: Hampton Press.

Yoo, Kyung Hyan and Gretzel, Ulrike (2008) 'What Motivates Consumers to Write Online Travel Reviews?', *Information Technology & Tourism*, 10.4, pp. 283–95.

Zelizer, Barbie (2004) 'When Facts, Truth and Reality are God-terms: On Journalism's Uneasy Place in Cultural Studies', *Communication and Critical/Cultural Studies*, 1.1, pp. 100–119.

7

Have Traveled, Will Write: User-Generated Content and New Travel Journalism

Usha Raman and Divya Choudary

Introduction

Open any weekend newspaper and you will see at least one travel feature, often written by a freelancer whose passion for travel is equaled by enthusiasm for talking about it. On the Internet, thousands of sites celebrate and promote travel, explicitly through exhortations to 'come and visit' and glowing accounts of facilities and services associated with a place. Blogs are supplemented by 'likes' on Facebook and comments on travel sites; homespun photos and videos compete for attention with professional attempts to showcase destinations. Another layer is contributed by amateur reviews and ratings of hotels and restaurants, tourist attractions and other travel services.

Together, these different forms build a library of information by a diverse group of writers, both professional and amateur. From the choice of where to go on vacation to choosing the airline, hotels and local tour operators, travelers depend increasingly on information that is provided by 'people like us', unreviewed and unscrutinized by a trained editorial eye. Writing about leisure travel is an increasingly routine exercise for certain people. Their texts set in motion a specific understanding of travel as a practice and as an act of consumption.

Travel journalism and writing has been gaining visibility in India, with more Indians traveling both within the country and abroad. While India has always had a certain exotic appeal for Western tourists, domestic and international travel by Indians has seen a spurt following the post-liberalization era, beginning in the late 1980s. In 2012, while the number of foreign tourist arrivals in India increased by 4.3 per cent, the increase in the departures of Indian nationals was

6.7 per cent. Within the country, domestic tourist visits grew by 19.9 per cent (Ministry of Tourism, 2013). Vacationing is becoming the norm among the growing middle class, with exploratory travel by younger Indians becoming more acceptable. Concomitantly, more Indians are sharing their travel experiences across a wide range of forums – online, in special- and general-interest magazines, in newspaper feature supplements and on television travel shows.

This chapter examines a subset of amateur travel writing in India on blogs and in newspapers. Our aim is to understand the intent and motivations of the writers, and the ways in which these contributions create certain ideas about the enterprise of tourism, the idea of travel and the making of travel memories.

The writing of travel, online and off

In recent years, travel writing – both journalistic and other forms – has attracted increasing attention as a legitimate and important form of discourse. Fürsich and Kavoori (2001) emphasize its importance as both evidence of and contributor to media globalization and the production of modern, global citizens. Speaking of 'tourists with typewriters', Holland and Huggan (2000, p. vii), quoting Fussel, point out that while travel writing has been criticized as a genre for 'second rate literary talents', it is also one of the most widely read literary forms today.

From backpacking students to the ultra-elite jetsetters, the number of travelers of all descriptions from emerging economies is fast exceeding in actual numbers those from traditional tourism markets. This is accompanied by a parallel increase in media consumption in these countries. For instance, in India the number of television channels dedicated to lifestyle themes is mirrored in the rise in special-interest lifestyle features in print media. The Indian newspaper *DNA* recently noted that travel shows were the 'hottest thing' on Indian television (Nandy, 2011). However, there is relatively little analysis of travel writing from postcolonial India outside of literary studies, a gap that needs addressing, given the growing importance of this sector in the emerging economies. Fürsich and Kavoori (2001, p. 162) argue that it is important to ask how 'new constituencies' of travel writers frame their experiences.

As the appetite for travel news and information, and travelers' accounts grows, traditional media outlets have found it difficult to invest resources in this form of journalism. However, more and more people engage in leisure travel, and many of them want to tell their stories. Hill-James (2006, p. 3) speaks of the 'citizen tourists' who 'engage in private experience but enact it in the public' through their writings. While large newspapers have staff reporters and itinerant journalists from other beats to provide content for their travel columns and only sometimes take contributions from amateur writers,[1] in India small newspapers depend on amateur contributors and professional freelancers for their travel content. Thus, many of the stories in these newspaper sections are written by non-professionals, those who travel for pleasure and write, in a sense, to extend and share that pleasure. These contributors do not intend to make a living from writing; their stories resemble extended journal entries, focusing on individual experience, the structure unmindful of journalistic conventions. These 'amateur' expressions deserve to be analyzed as alternative journalism, in line with Atton and Hamilton's (2008, p. 2) suggestion to include the 'journalisms ... of popular culture and the everyday', even if they are not immediately driven by social empowerment.

Apart from the continuing popularity of travel in the mass media, many seek information on the Internet. Gretzel and Yoo (2008) mention that searching for travel-related information is one of the most popular activities online. The veritable sea of travel information on the web includes websites of tourism companies, portals of tourism bureaus and travel magazines, and information and opinion from travelers. As of June 2013, TripAdvisor featured over 100 million travel reviews and over 14 million candid traveler photos (TripAdvisor.com, 2013) – a significant increase from the 10 million travel reviews and 750,000 photos that Yoo and Gretzel (2008) had noted in September 2007.

How do these two distinct but related forms of what we might call 'user-generated' content (newspaper features contributed by casual travelers and online narratives such as travel blogs) fit into the larger discourse on travel, particularly in an emerging economy such as India? What do they mean to both the writers ('content producers') and readers or users? How does this discourse relate to the context of travel and tourism in India?

Drawing upon an analysis of material generated by Indian amateur travel writers, we argue that the current economics of travel journalism and the independence offered by online media combine to create a space for a more interactive culture of information sharing among travelers. This form of travel writing also consciously sets itself apart from the market forces that tend to influence lifestyle reporting in mainstream media, and takes on an individualistic, even idiosyncratic point of view.

Content from 'real people'

The close relationship between the tourism industry and travel writers has been much discussed (Britton, 1979; Fürsich, 2002), generating questions of authenticity and bias. Readers and viewers are suspicious of such 'underwritten' content and prefer to look to independent accounts for their information needs. Many scholars, notably Yoo et al. (2009) and Gretzel and Yoo (2008, p. 38), have noted that non-commercial media are more likely to be trusted by users; they have also found that (travel) review readers are highly educated, have high incomes, travel frequently and are heavy Internet users. There is a large overlap between those who write travel pieces, both online and offline (in traditional print media), and those who consume such writing. Going in search of the 'authentic' experience of 'people like me', a prospective traveler is more likely to seek out user-generated accounts. About 80 per cent of travelers worldwide find user reviews important (Statistic Brain Research Institute, 2013).

The social web has blurred the distinctions between the public and private, creating a climate where individuals are encouraged to participate in an ongoing conversation about their experiences and share their opinions in a variety of forums. Writing about leisure travel increasingly finds its way into these spaces; individuals generate 'narbs' or 'narrative bits' (Mitra, 2010) that together tell evocative – and in some readings, authentic – stories. One category of user-generated content that has been studied includes consumer reviews on travel planning sites, such as TripAdvisor and SkyTrax. Some scholars have characterized such user narratives as 'inert' representations of journeys undertaken (Bissell, 2012, p. 152). Such content has a certain instrumentality and immediacy of purpose – to provide quick and honest feedback to fellow travelers about specific services. In contrast, blogs and feature stories allow the writer to engage with the experience of travel in a variety of ways.

Fürsich and Kavoori (2001, p. 157) earlier questioned the tension between 'the textualization of authenticity and the actual experience' of travel, which to some extent is relieved in 'unmediated' writing of the kind found in blogs. In an analysis of 30 travel blogs, Bosangit, McCabe and Hibbert (2009) find that bloggers' narratives provided important insights into identity construction and sense-making among tourists (both writers and their readers). While this is particularly relevant for online writing, it is also true of contributory travel writing in newspapers and magazines.

It is also important to recognize that in today's media ecology, there is a seamless and overlapping consumption and production of various forms of content: writing by professionals, non-professionals, users and service providers, experts and amateurs often overlaps in a palimpsest of texts. Citizens write and *write into* the news, users create, comment and edit (as on Wikipedia). They promote their work through other media, such as Twitter and Facebook, strategically drawing in other consumers to their expressive products. Castells (1996), in his commentary on the new media landscape, alludes to the idea of 'prosumers' (a term originally coined by Alvin Toffler), referring to those who actively participate in the act of creating culture (and content) while also consuming it, but in doing so, also further the cause of the market or the sphere of production of which they are also users. Slightly different is the notion of 'produser', a productive user who adds to the knowledge base independent of commercial producers, signifying a process that is predicated on the collaborative content creation spaces offered by Web 2.0 (Bruns, 2008, p. 10). While travel bloggers may not see themselves necessarily as knowledge producers in Bruns' definition, they do contribute to the economy and practice of travel in significant ways. Electronic Word of Mouth (EWOM) or 'word of mouse' (Volo, 2010) has 'dramatically expanded the scope for citizens to engage in practices of public communication that are synonymous with journalism' (Swift and Nitins, 2011, p. 8). Pecquerie and Kilman (2007, p. 3) similarly note that journalism is no longer a 'specialized profession with a unique set of rules and ethics' and that bloggers are 'complementary content producers'.

In the content flow that is travel writing, this 'complementary content' acquires a powerful and distinct voice, seen as free of commercial influence and preferential or limiting editorial gatekeeping. Other sectors of commerce and industry have already seen the impact of user-generated content on the flow of opinion and information on the uptake and health of brands (Dhar and Chang, 2009; Reigner, 2007). There is reason to believe that this is true of the travel trade, too. Whether it is the contributions of enthusiastic travelers eager to share experiences and provide advice in the columns of a newspaper, or compulsive bloggers who set out to recreate their journeys and in the process create circles of reader-consumers, this content has the ability to shape the discourse and politics around tourism and travel.

Deciphering user-generated content

In order to explore user-generated travel content in India, we focused on two specific sites: two leading newspapers and two travel blogs. Newspaper travel features published between 15 December 2012 and

15 January 2013 were selected from *The Times of India* and *The Hindu*, two of the most widely read English language papers in India (Media Research Users Council, 2012).[2] This time of year coincides with school holidays, and a concomitant rise in travel and travel-related articles. *The Times of India* is known to make the effort to reach younger readers, while *The Hindu* has a reputation for being independent and reliable. Both papers offer online versions, which allow access to readers' comments on the articles.

We considered only those features contributed by casual travelers as user-generated content. Articles written by staff reporters, contributed by news agencies, or without bylines were not included for analysis. As neither newspaper distinguished between the staff writers and freelance or amateur contributors in terms of bylines, we used Google and LinkedIn profiles to verify whether the writers were on staff.

In *The Hindu*, travel features were published mainly in the daily feature supplement, 'Metroplus' and the 'Sunday Magazine'. Of the 52 travel-related articles that were published during the review period, 11 features (around 21 per cent) qualified as user-generated content by our definition.

In *The Times of India*, travel features are part of the 'Life & Style' section. In the month of review, 45 travel-related articles were found, of which only four (around nine per cent) fit our criteria.

The travel blogosphere – even its Indian subset – is a large place to traverse. Blogs vary in style, content and form. Our purpose was to attempt to position blogging within the larger discourse on travel. After a general search, we used websites like Quora, blogRank and hoteldepot that listed popular Indian travel blogs according to two parameters – number of followers and page views. Of the top five blogs written by Indians and with a large number of entries, the three most popular ones were written by women travelers, and the two that were finally selected for analysis – based on the volume of text, balance of photographs and visuals, and range of sites featured – were The Shooting Star (http://the-shooting-star.com/travels/) by Shivya Nath on Wordpress (Alexa rank 244,944) and Anuradha Goyal's travels (http://anuradhagoyal.blogspot.in/) on Blogspot (Alexa rank 442,779). In our analysis, most of the writers – feature writers and bloggers – were found to be women. Within the top 30 travel blogs in 2013, 60 per cent are by women, and of the newspapers features looked at, 10 of the 16 were by women (Hoteldepot, 2013). This is in line with earlier findings that show that most travel journalists are women (Hanusch, 2012).

I travel, I write, I become

Newspapers

A cursory reading of features by staff reporters reveals that their content is very prescriptive and informative, intended for prospective travelers with a possible goal of promoting tourism and culture. The travel features contributed by readers, in *The Times of India*, are not very different from those by staffers, being similarly descriptive and factual in nature as seen in this extract:

> In Santa Park, you can visit Santa's cavern, ... and meet Santa and have your photo taken with him. Gifts and souvenirs are available at the village, and if you'd like to be close to the holiday cheer, hire a cabin at the village. (Nazareth, 23 December 2012)

The Hindu travel features are often about more than just travel – for instance, a trekker's account of a tryst with butterflies in a sanctuary (Padmanabhan, 29 December 2012), a traveler's religious experience (Jalil, 13 January 2013), or a travel feature about the evolving food culture and finding home food in foreign lands:

> A culture that once varied between formal dining and cold packed lunches, with little in between, now finds itself embracing an altogether different style of eating. ... The quality of what's on offer must not be disparaged; al fresco dining in winter needs to be worth the consequential numb fingers, and London, in its own way, is ensuring that it's certainly worth it. (Promod, 22 December 2012)

While our intent was not to compare the two newspapers, the difference in the type of travel content published serves to demonstrate the difference in the positioning of 'travel and tourism' within each paper, and the audience to which each newspaper caters. *The Hindu* is generally seen as a more conservative, less commercial paper, with an audience that fits this profile – an older demographic, interested in more 'serious' content. *The Times of India*, however, positions itself as a paper for young people on the move, urban, trendy and westernized. *The Times of India* also has been criticized for emphasizing the financial bottom line, hence building bridges with advertisers through conscious content placement (Brown, 2005). In *The Hindu*, therefore, the travel writing tends to be more experiential, with writers playing the role of cultural interlocutors, as seen in the excerpt from the article on food culture. *The Times of India* features tend to focus on the 'doing' and 'seeing' aspects of travel as demonstrated earlier in the excerpt from the article on the Santa Claus village.

Both newspapers have a web presence, encouraging reader participation through comments and sharing on social media. However, most of the articles had not received any substantive comments on the website. This may also point to a more basic difference between newspaper content (even when online) and blogs, which are seen as interactive spaces that invite participation from the users. In newspapers, amateur writing is not particularly distinct from the rest of the content. Thus, it is likely that they are not perceived differently by readers.

Blogs

A close reading of the blogs included the visual and textual elements, the pattern of labeling as evidenced in the tag clouds, style of writing, content and the bloggers' self-description. We also paid attention to the comment thread, which provides a sense of the community (or 'audience') that is generated by the blog.

The Shooting Star blogger Shivya Nath quit her corporate job in 2011 to 'be able to travel as and when [she] pleased'. Information shared on her blog indicates that she has been to more than 20 countries, has co-founded a 'responsible travel' venture, works as a freelance social media consultant, and writes for many online and offline publications. The blog lists her travel plans, and links to publications in which her work has appeared.

The initial entries in her blog, created in April 2008, were musings about her work and life, book and food reviews, and it was only in November 2010 that she began to blog about her travels. This quotation gives a sense of her tone:

> Sometimes I think we can measure our lives by the places we've seen. There's so much beauty in the world, so many undiscovered gems, and yet so many of us will spend our days in a small little box, unable to escape the vicious work-money cycle ... (3 December 2010).

The next set of entries were prescriptive in nature, such as '5 Weekend Getaways from Singapore to Pamper You.' After Nath (10 April 2011) launched the Incredible India Social Experiment with the aim to 'collaborate and showcase India's richness of travel offerings online, facilitating both domestic and international tourism in India', her blog took on a more serious tone – with posts of travel tales, experiences and impressions of places, and tips for prospective travelers. In September 2011, she shared her first published travelogue 'Of hitch-hiking in India', that had also been published on *Clay* – Club Mahindra's travel portal (CLAY, 2011). Her work has since been published in newspapers like *The Times of India* and *The Hindu* and in several travel magazines.

Nath's published work caters to the style of the publication. For instance, in an article in *The Times of India* (Nath, 8 February 2013) she shares a factual account of her experience and information about the destination. Her article published in *The Hindu* (Nath, 23 February 2013) talks of her feelings as well – 'When we part ways for the night, I carry with me a strange sense of longing; I long for their contentment, the innocence of their thoughts, and the simplicity with which they live in this little green haven.'

Her posts are about her overall experience of traveling, right from the planning to the actual journey, her expectations, disappointments and reflections on travel. Tips for travelers include must-dos, inspirations and safety measures. In her post 'All You Need is a Backpack & a Heart For Adventure', Nath writes:

> Growing up in India, many of us have associated traveling with luxury, shopping (for the mothers), and religious getaways. We've always been taught to look at travel as a holiday from life, not as life itself. In fact, we haven't learnt to be travelers, just tourists at best .... Luckily, times are a-changing. ... I for one, have bombarded myself with blogs & tweets of people who are out to discover the world. (15 September 2011)

Apart from posting comments responding to the entries or raising questions about solo travel and funds, readers (from across the world) often provide further details and recommendations.

In her blog, Nath often writes about less popular destinations, places that are untouched by tourism and 'crowds', and this is clear from the largest tag cloud 'Offbeat', with 60 posts. The other tag cloud categories are India (55), Weekend Getaways (39), Culture (33), Budget (29), Travel Inspiration (29) and Adventure (27). The most popular posts on her blog are ones that offer tips – '10 Must-Try Vegetarian Food Places in Singapore' and '9 Countries That Offer Visa on Arrival For Indians'.

Anuradha Goyal Travels blogger Anuradha Goyal, who calls herself an 'eternal nomad', began blogging in 2004, sharing her experiences while traveling and her reviews of the books she was reading. Travel writing remains her hobby, while consulting in business innovation is her main source of income. In the 'About' section of her blog, Goyal admits that there is a 'yearning to explore the world first hand that keeps [her] going'. Although some of her work has been published in *The Hindu*

(Goyal, 23 February 2013), she makes no mention of this on her home page or even in the 'About' section.

Goyal's initial posts included musings and thoughts, and unlike other popular travel blogs, she continues to post these. Most of the popular posts feature her experiences in India as emphasized in the tag clouds categories, with the largest being Travel, Bangalore, Delhi, Opinion and Ancient India.

Goyal interacts with officials, tourists and locals to get a better sense of the place she is in. Her posts talk about folk art, local cuisines, religion, culture, tribes, heritage sites, temples and architecture and are rich in details. For instance, she shares with readers a conversation she had with a man about wearing shoes inside a church prompting her to ask: 'Should you treat God the way that God's religion demands or should you treat each God like the one you believe in?' (27 April 2013).

Goyal often narrates the stories of people she has met on her travels. About a World War II soldier she met in Poland who carried with him a photo album with pictures of himself on a horse in various places across Europe, she writes, 'I took pictures of his pictures to be able to tell his story to you' (2 April 2013).

The comments on her posts indicate that her readers enjoy the detail, often wanting more. On the post about the soldier (2 April 2013), a reader comments: 'What he told you about the war and post war? I'm wondering what were all those badges for [sic].' Unlike in Shivya Nath's blog, most of the comments on this site were from Indian readers.

Writing to break free

The difference in the content and style of writing indicates that travel blogs follow no set rules or specifications, unlike newspapers that need to cater to the taste of their audience and work within an editorial framework. Some bloggers like Anuradha Goyal, have a narrative style of writing and an interest in story-telling, providing detail meant for armchair travelers. Others like Shivya Nath write to share experience and also to address the needs of travelers who not only want to sit back and read but are ready to go on a trip.

A significant way in which the newspaper features differed from the blogs is in the direct expression of opinion – from the experience of travel to the food, cost, service and quality of transport. The newspaper features are much more positive, hardly ever critical of services or facilities. Travel journalism in general has been found to be largely positive, perhaps because those who write about 'foreign' locations also see themselves as ambassadors for these cultures (Austin, 1999;

Hanusch, 2012). For instance, in a newspaper feature, what some might see as crowded, dirty and noisy streets becomes bustling, colorful and lively:

> They streamed down side streets and lanes and little winding alleys, poured into the broad Grand Road till it became a flowing river of humanity. And still more surged in, packed tighter and tighter till all we could see ... was a sea of heads streaked with snaking currents of saffron, red and white as sects and cults merged with flood of devotees. (Ganzter and Ganzter, 23 December 2012)

Bloggers have more freedom to follow their own style, with no gate-keepers to regulate the content. Criticism of locations, people and hospitality or tourism establishments, the ability to question stereotypes and freedom to depict places and events allow the blogger to be more 'honest'. For instance, blogger Shivya Nath in her post on Paris writes:

> I grab an ice-cream to cool off from the crowds, and walk to the Eiffel Tower, only to be greeted by more queues & countless Eiffel Tower tours. I fulfill the customary photo-taking criterion and walk away. Just like that, my grand illusions of Paris are crushed one by one .... My inability to speak the local language adds fuel to my fire, as time and again, many locals look cross for trying to initiate a conversation in English. I begin to think of Paris as a hugely successful PR campaign; if you don't enjoy sightseeing in Paris, you are not classy or sophisticated enough. (11 October 2011)

The free interaction between the blogger or generator of content and the readers is once again demonstrated in the comments following this post:

> This is a very interesting post ... and goes against all the stereotypes we set for ourselves. On my Eurotrip, I didn't enjoy Paris much ... it was crowded, touristy and somehow, very alien – whereas I absolutely loved Berlin, a city with definitely not a romantic image like Paris. (11 October 2011)

While another commentator writes:

> I think the key to Paris is to explore the city itself ... So please, please, PLEASE give Paris another chance – it's one of my favorite cities in the world and has a quiet splendor like no other! (11 October 2011)

One might expect that the medium affects both the production and reception of texts, and writing in conventional print media, even when produced by non-professionals, or consumers, would differ from the narratives on blogs. We found that newspaper features tend to celebrate the global citizen of the world, the sophisticated traveler and holiday-seeker, and are often physically positioned within a lifestyle supplement alongside advertisements of holiday packages. Even user-generated content in conventional print publications, then, is framed within the conventions of the medium. On the other hand, one might see the blog as a somehow solitary journey, in terms of both the reading experience online and its presentation on screen, an isolated stream of text with nothing to distract from its content unless the reader decides to switch windows. In contrast, the travel article in a newspaper must – because of the physical limitations of its size – position itself within yet apart from the busy-ness and humdrum of daily life, hemmed in by politics, social issues and corporate activity.

In contrast, the bloggers position themselves outside of mass tourism. They are travelers, who reject the stereotype of the tourist. As travel writer Dalrymple argues:

> [T]ravellers tend by their very natures to be rebels and outcasts and misfits: far from being an act of cultural imperialism, setting out alone and vulnerable on the road is often an expression of rejection of home and an embrace of the other. (Dalrymple, 2009)

Considering Cohen's influential tourist role typology (1972), the bloggers see themselves more as 'explorers' and 'drifters', clearly connecting travel to a search for meaning. In contrast to what Dalrymple suggests, however, they do not feel the need to reject one home in order to embrace another. As Nath suggests, 'maybe we are indeed the sum of the places we travel to' (23 September 2012).The search for difference is temporary. In her post 'Gargnano; Life or Something Like it in Italy', she writes:

> With the onset of darkness, the visitors start to disappear into the far shores of the lake and I return to what feels like my own little village. Someday, I must return to live more of this simple, slow and beautiful life. (23 May 2011)

The bloggers' experiences go beyond the 'recreational', moving toward 'experimental, experiential and existential' modes. At many points in the two bloggers' narratives, it is possible to discern a search for inner meaning, in a sense 'going away' to 'find oneself'. While Goyal admits that the 'yearning to explore the world first hand [is what] keeps [her] going', in her post 'Meri Mitti Ki Sugandh – Yearning for my own culture', she considers the strong ties to her homeland:

> Does this mean my homeland and my mother tongue are such an integral part of me, that I remember them only when I miss them. What does my own culture mean to people like me, who have grown up in different parts of the country and have seen a few cultures outside the country? Does it help me appreciate and understand my own roots in [sic] much better way? (23 November 2004)

Both Nath and Goyal are 'untourists' (Corrigan, 1997), who distance themselves from the enterprise of tourism. Clearly, they are part of an elite, who can employ travel as a means of (self) discovery, but in doing so, also allow Internet users to find ways of escaping the commercial web of tourism.

Filho and Tan (2009) explain that user-generated content is a means of consumer empowerment in the travel industry, with electronic word-of-mouth serving as a powerful antidote to the hard sell of travel agencies and tour operators. This also echoed Gretzel and Yoo's results of a 2008 web-based survey where 97 per cent of Internet users said they read online reviews while planning a trip, which the researchers interpret as an indication of the 'democratization' of travel and, we would add, of the production of travel information. Swift and Nitins (2011) similarly found that blogs were among the top three online platforms for seekers of travel-related information. The bloggers indicate that they are 'in charge' of their travel plans and offer readers a way to gain control of their itineraries as well. The travel articles in newspapers, while offering a personal viewpoint, speak less directly to the planning process, a function of both the unidirectionality of the medium and the location of the writing within the more 'formal' structure of professional media.

When looking at travel blogging as a productive practice, one might see the travel blogger as one who is reclaiming the world on his or her own terms. For the Indian travel blogger, writing about the travel experience provides a way to create and claim a realm of experience that is both geographical and cultural, yet one that is denied to her in the conventional discourse of commodified tourism and travel. The visual narrative of travel promotion is decidedly in favor of group and family travel as safe experiences even where the terrain is rugged and wild. The solitary female travel bloggers are able to pull out an individualized experience of travel that takes it out of the packaging of the tourism industry. They shape their experience in words and pictures of their choosing, turning their gaze outward (with their lens) and inward (through their words) to an audience that exists only in the confines of their blog. Writing blogs to 'give back to the travel community' confirms Swift and Nitins' finding (2011) that the top three motivators of bloggers are to have a visible travel history, to promote and share travel plans, and to receive feedback on their travel stories.

Our readings of the Indian travel blogs suggests that Indians who travel, and who write about travel, do not necessarily see themselves as 'Indians who travel' but rather position themselves as global citizens who have the freedom and the ability to give themselves the experience of different places and cultures. This does not always make them any less political, as one can see in the writings of Shivya Nath, nor are they less conscious of their cultural roots, but they are at ease in the world at large.

Our analysis represents an initial attempt to position travel blogs and amateur travel writing in the print media within the larger discourse on travel, and bringing them within the purview of research on this genre. It is also an attempt to bring in non-Western, yet 'global' texts into a space of study that has largely concentrated on travel writing from the global North. Admittedly, there are limitations that can be addressed as work in this area is taken forward and put in context with other studies on user-generated content and travel writing, and related to our understanding of how people create and use content across media.

With travel articles in newspapers, the newspapers themselves and the market forces to which they cater have a homogenizing influence on the content, whether written by amateurs or professionals. Given that 'writing in' to a newspaper takes more effort and commitment, comments in this medium are extremely rare. The blog, combined with the circle of its readers/followers, represents a community of travel practice, creating and sustaining a discourse that is both *instrumental* (including discrete references to issues of 'what' and 'how' of travel) and *transcendental* (emphasizing discovery and distance from the ordinary or mundane). In some ways, blogging is an extension of other forms of travel journalism and travel writing, particularly in the role of information sharing and advice disseminating. In other ways, as in its open-endedness, its ability to sustain a reflective tone, and its potential to build communities of interest and practice outside the commercialism of the industry, travel blogging represents a significant departure from mainstream professional travel journalism.

Notes

1. S.B. Vijaya Mary (telephone interview, 18 April 2013) is the senior assistant editor at *The Hindu Metroplus*.
2. According to the Indian Readership Survey, *The Times of India* has registered an *average issue readership (AIR)* of 7,643,000 in Q2 while *The Hindu* registered an AIR of 2,208,000 in the same quarter. As two of the oldest and the largest newspapers in the country, they are considered the newspapers of record and serve the 'national' audience.

References

Atton, Chris and Hamilton, James F. (2008) *Alternative Journalism*, London: Sage.

Austin Elizabeth (1999). 'All expenses paid: Exploring the ethical swamp of travel writing' *The Washington Monthly*, 31.7. Available at: http://www.washington-monthly.com/features/1999/9907.austin.expenses.html.

Bissell, David (2012) 'Mobile testimony in the information age: the powers of travel reviews', *International Journal of Cultural Studies*, 15.2, pp. 149–64.

Bosangit, Carmela, McCabe, Scott and Hibbert, Sally (2009) 'What is Told in Travel Blogs? Exploring Travel Blogs for Consumer Narrative Analysis', in Wolfram Höpken, Ulrike Gretzel and Rob Law (eds), *Information and Communication Technologies in Tourism 2009*. New York: Springer, pp. 61–71.

Britton, Robert A. (1979) 'The image of the Third World in tourism marketing', *Annals of Tourism Research*, 6.3, pp. 318–29.

Brown, Robert (2005) 'The importance of being earnest as well as entertaining', in Nalini Rajan (ed.), *Practising Journalism: Values, Constraints, Implications*. New Delhi: Sage, pp. 242–54.

Bruns, Axel (2008) *Blogs, Wikipedia, Second Life and Beyond: From Production to Produsage*, New York: Peter Lang.

Castells, Manuel (1996) *The Rise of the Network Society: The Information Age: Economy, Society and Culture, Volume 1*, Cambridge, MA: Blackwell.

CLAY–Club Mahindra Travel Experiences (2011) *Spiti: The India that lives in a cold mountain desert*. Available at: http://www.clubmahindrablog.com/spiti-the-india-that-lives-in-a-cold-mountain-desert.

Cohen, Erik (1972) 'Toward a sociology of international tourism', *Social Research*, 39.1, pp. 164–82.

Corrigan, Peter. (1997) *The Sociology of Consumption: An Introduction*, London: Sage.

Dalrymple, William (2009) 'Home truths on abroad', *The Guardian*, 19 September. Available at: http://www.guardian.co.uk/books/2009/sep/19/travel-writing-writers-future).

Dhar, Vasant and Chang, Elaine A. (2009) 'Does chatter matter? The impact of user-generated content on music sales', *Journal of Interactive Marketing*, 23.4, pp. 300–7.

Filho, Luiz M. and Tan, Felix B. (2009) 'User-Generated Content and Consumer Empowerment in the Travel Industry: A Uses & Gratifications and Dual-process

Conceptualization', *Pacific Asia Conference on Information Systems PACIS 2009 Proceedings*. Paper 28. Available at: http://aisel.aisnet.org/pacis2009/28.

Fürsich, Elfriede (2002) 'Packaging culture: the potential and limitations of travel programs on global television', *Communication Quarterly*, 50.2, pp. 204–26.

Fürsich, Elfriede and Kavoori, Anandam P. (2001) 'Mapping a critical framework for the study of travel journalism', *International Journal of Cultural Studies*, 4.2, pp. 149–71.

Ganzter, Hugh and Ganzter, Coleen (2012) 'Chariots of the gods', *The Times of India*, 23 December. Available at: http://articles.timesofindia.indiatimes.com/2012-12-23/travel/35969641_1_lord-jagannath-balabhadra-gundicha-temple.

Goyal, Anuradha (2004) 'Meri Mitti Ki Sugandh – Yearning for my own culture' *Anuradha Goyal Travels* [Blog], 23 November. Available at: http://anuradhagoyal.blogspot.in/2004/11/meri-mitti-ki-sugandh-yearning-for-my.html.

Goyal, Anuradha (2013) 'Ski and Beyond', *The Hindu*, 23 February. Available at: http://www.thehindu.com/features/magazine/ski-and-beyond/article4438753.ece.

Goyal, Anuradha (2013) 'Travel Stories VI: Meeting the World War II Soldier', *Anuradha Goyal Travels* [Blog], 2 April. Available at: http://anuradhagoyal.blogspot.in/2013/04/WorldWarIISoldier.html.

Goyal, Anuradha (2013) 'Magical Mumbai V: Afghan Church, Colaba', *Anuradha Goyal Travels* [Blog], 27 April. Available at: http://anuradhagoyal.blogspot.in/2013/04/Afghan-Church.html.

Gretzel, Ulrike, and Yoo Kyung H. (2008) 'Use and Impact of Online Travel Reviews', in Peter O'Connor, Wolfram Höpken and Ulrike Gretzel (eds), *Information and Communication Technologies in Tourism 2008*. New York: Springer Wien, pp. 35–46.

Hanusch, Folker (2012) 'A profile of Australian travel journalists' professional views and ethical standards', *Journalism*, 13.5, pp. 668–86.

Hill-James, Candida R. (2006) Citizen Tourist: Newspaper Travel Journalism's Responsibility to its audience. Unpublished Master's thesis. Brisbane, Queensland University of Technology. Available at: http://eprints.qut.edu.au/16304/.

Holland, Patrick and Huggan, Graham (2000) *Tourists with typewriters: Critical Reflections on Contemporary Travel Writing*, Ann Arbor: University of Michigan Press.

Hoteldepot.in (2013) *Top Travel Bloglist 2013*. Available at: http://www.hoteldepot.in/blog/travel/30-top-travel-blogs-in-india-in-2013/.

Jalil, Rakhshanda (2013) 'A pilgrimage to remember', *The Hindu*, 13 January. Available at: www.thehindu.com/todays-paper/tp-features/tp-sundaymagazine/a-pilgrimage-to-remember/article4303282.ece.

Media Research Users Council (2012) *IRS 2012 Q2 Topline Findings*. Available at: http://mruc.net/images/irs2012q2-topline-findings.pdf.

Ministry of Tourism (2013) *India Tourism Statistics at a Glance 2012*. Available at: http://tourism.gov.in/writereaddata/CMSPagePicture/file/marketresearch/Ministry%20of%20tourism%20English%202013.pdf.

Mitra, Ananda (2010) 'Creating a presence on social networks via Narbs', *Global Media Journal*, 9.16, pp. 1–18.

Nandy, Priyadarshini (2011) 'Travel shows on India are the hottest thing on television', *DNA*, 10 August. Available at: http://www.dnaindia.com/entertainment/1574517/report-travel-shows-on-india-are-hottest-thing-on-television.

Nath, Shivya (2011) 'Incredible India: A social experiment in travel', *The Shooting Star* [Blog], 10 April. Available at: http://the-shooting-star.com/2011/04/10/incredible-india/.

Nath, Shivya (2011) 'Gargnano; Life or Something Like it in Italy', *The Shooting Star* [Blog], 23 May. Available at: http://the-shooting-star.com/2011/05/23/gargnano-lake-garda-travel-blog/.

Nath, Shivya (2011) 'All you need is a Backpack & a Heart For Adventure', *The Shooting Star* [Blog], 15 September. Available at: http://the-shooting-star.com/2011/09/15/travel-inspiration/.

Nath, Shivya (2011) 'Of Hitch-Hiking in India', *The Shooting Star* [Blog], 13 September. Available at: http://the-shooting-star.com/2011/09/13/hitch-hiking-india-travel-blog/.

Nath, Shivya (2011) 'Falling in love with Paris', *The Shooting Star* [Blog], 11 October. Available at: http://the-shooting-star.com/2011/10/11/falling-in-love-with-paris-travel-blog/#more-2380.

Nath, Shivya (2010) 'Once upon an island', *The Shooting Star* [Blog], 3 December. Available at: http://the-shooting-star.com/2010/11/03/once-upon-an-island/.

Nath, Shivya (2012) 'How Travelling is Breaking My Heart', *The Shooting Star* [Blog], 23 September. Available at: http://the-shooting-star.com/2012/09/23/how-travelling-is-breaking-my-heart/.

Nath, Shivya (2013) 'Go native in Ile Maurice', *The Times of India*, 8 February. Available at: http://articles.timesofindia.indiatimes.com/2013-02-08/travel/36721080_1_mauritius-ravi-boat.

Nath, Shivya (2013) 'Life as we knew it', *The Hindu*, 23 February. Available at: http://www.thehindu.com/features/magazine/life-as-we-knew-it/article4438743.ece.

Nazareth, Sonia (2012) 'Visiting the Santa Claus village in Lapland', *The Times of India*, 23 December. Available at: http://articles.timesofindia.indiatimes.com/2012-12-23/travel/35982073_1_santa-claus-village-korvatunturi-finnish-lapland

Padmanabhan, Sujatha (2012) 'The butterfly effect', *The Hindu*, 29 December. Available at: www.thehindu.com/features/magazine/the-butterfly-effect/article4252724.ece.

Pecquerie, Bertrand and Kilman, Larry (2007) 'From Citizen Journalism To User-Generated Content – Establishment media organizations with user-generated content', *eJournal USA*. Available at: http://www.america.gov/st/democracyhr-english/2008/November/20080518181554WRybakcuH0.2765467.html.

Promod, Sheba (2012) 'Hot food on a cold day', *The Hindu*, 22 December. Available at: www.thehindu.com/todays-paper/tp-features/tp-sundaymaga-zine/hot-food-on-a-coldday/article4230635.ece.

Reigner, Cate (2007) 'Word of mouth on the web: the impact of Web 2.0 on consumer purchase decisions', *Journal of Advertising Research*, 47, pp. 436–47. Available at: http://instruct.uwo.ca/mit/3771-001/Word_of_Mouth_On_the_Web__Cate_Riegner.pdf.

Statistic Brain Research Institute (2013) *Internet Travel Hotel Booking Statistics*. Available at: http://www.statisticbrain.com/internet-travel-hotel-booking-statistics/.

Swift, Adam G. and Nitins, Tanya (2011) 'Building engaged, sustainable online communities: a case study of an adventure travel website', in Yahya R. Kamalipour (ed.), *Proceedings of the 5th GCA Conference: Global Power Shifts: Impact on Economy, Politics, Culture and Media*, Kuala Lumpur.

TripAdvisor (2013) *Fact Sheet*. Available at: http://www.tripadvisor.com/PressCenter-c4-Fact_Sheet.html.

Volo, Serena (2010) 'Bloggers' reported tourist experiences: Their utility as a tourism data source and their effect on prospective tourists', *Journal of Vacation Marketing*, 16.4, pp. 297–311.

Yoo, Kyung-Hyan, Lee, Yoonjung, Gretzel, Ulrike and Fesenmaier, Daniel, R. (2009) 'Trust in Travel-related Consumer Generated Media', in Wolfram Höpken, Ulrike Gretzel and Rob Law (eds), *Information and Communication Technologies in Tourism 2009*. New York: Springer, pp. 49–59.

Yoo, Kyung H. and Gretzel, Ulrike (2008) 'What motivates consumers to write online travel reviews?', *Information Technology & Tourism*, 10.4, pp. 283–95.

8
Going with the Flow: Chinese Travel Journalism in Transition

Jiannu Bao

Introduction

This chapter explores the evolution of travel journalism in the context of unprecedented institutional and technological changes of the Chinese media system and a boom in tourism development over the past three decades. Economic reforms since 1978 resulted in considerable increases in disposable incomes, which led to a wave of consumption, including leisure travel, which has swept across both urban and rural China since the early 1980s. Increases in leisure time – following the institution of the two-day weekend in 1994 and annual three one-week vacations in 1999 – have made the experience of leisure travel even more desirable for the general public. Interest in new and unique ways of accessing travel information have inspired, and also forced media institutions to seize the opportunities presented in the great transition of media in China. They had to develop new formats and popular content to increase advertising and circulation revenues after the government cut down its direct funding for most media outlets except a few major institutions in 1992. Travel journalism has since developed rapidly, first in the print and then in the broadcast media, and more recently on the Internet. The genre has become an important area for various types of media, which I categorize as the official, the negotiated and the flexible media.

This chapter details how travel journalism, with diversified formats and content, has become a recognized genre of popular and lifestyle journalism in various media types in China. It identifies a three-fold expansion model of travel journalism in China and distinguishes discourses of travel and tourism by media institutions of different categories. Finally, the issue of national identity is discussed in connection with travel journalism representation of the Other. It is a summary of a more complex project on the historic, economic and cultural aspects of travel media in China (Bao, 2005).[1]

Development of travel media in China

After the founding of the People's Republic in 1949, there was little media coverage of travel in China for almost three decades because freedom of mobility was restricted and travel was discouraged. Chinese media institutions followed the communist media model as propaganda machines used to publicize and educate the people in the party's policies (Lee, 2000).

Tourism and travel became a popular topic in the news media only as tourism started to take off after 1978. The media followed the government rhetoric and saw tourism expansion as a 'logical component of its post-Cultural Revolution normalization strategy' (Richter, 1989, p. 26), which could help – as the Party organ *People's Daily* declared in 1979 – 'promote mutual understanding and friendship' between China and other countries on one hand, and 'accumulate funds for the splendid plan of the Four Modernizations' on the other (*People's Daily*, 19 February 1979, p. 1). Inbound tourism became the predecessor for domestic tourism development.

As domestic tourism emerged, newspapers, especially evening papers, started to publish travelogues or personal accounts of travel experiences. Meanwhile, travel magazines sprang up, offering information about scenic spots and historic sites, cultures and customs of different places, and tour packages. Such information was helpful to tourists at a time when accurate travel information was not easy to come by.

Coverage of tourism and travel also increased in the broadcast media as television became a mass medium in 1980s. However, consumer-oriented guides to travel were rarely televised at that time. Instead, documentaries on scenic sites predominated.

It was not until the 1990s that travel journalism started to specialize and diversify as travel consumption grew rapidly. Chinese media institutions, which were compelled to turn to the market in this period, were quick to explore the emerging niche market, as consumers needed detailed information about travel and travel operators required information about travel demand. Some industry insiders even considered this era to be the 'golden period for the travel media' (Chen, 2013, p. 1).

Newspaper travel sections and travel magazines

Newspapers started travel sections for the 'dissemination of clustered information about travel' (Li and Xiao, 2003). The first was *Guangzhou Daily*, a party newspaper and a pioneer in Chinese media reform. It launched its travel section in 1992, followed by two outlets with official background – *Yangcheng Evening News* in the same city in 1993 and *Wenhui Daily*, a Shanghai-based culture-oriented broadsheet, in 1996. More newspapers followed suit. As insiders of these two publications describe, newspapers presented largely a mixture of information – tourism policies, trends and consumer guides (Hu, personal interview, 2003; Li, personal interview, 2003). However, in the late 1990s, many of them started to focus on serving individual travelers. *Wenhui Daily*'s travel section, for instance, turned to exploring new places and presenting new discoveries of its staff reporters. As the then-executive editor of the travel section, Hu Peijiong, explained that the goal was to 'offer something which you can never find in books' (Hu, personal interview, 2003). This strategy triggered even a new style of writing which mixed journalistic elements with personal experiences (Hu, personal interview, 2003). Mastering eyewitness reporting became a professional requirement.

Travel magazines flourished during the same period. Of dozens of travel magazines published throughout the 1990s, the three most popular have been operating under different business models. Since its establishment in 1995, *Traveler* (lvxingjia) has used the publishing license number of an internally circulated pamphlet compiled by the China Travel Service (CTS) Head Office. *Trends Traveler Magazine* (shishang lvyou), in Beijing, is published by the *Trends* (shishang) Media Group in association with International Data Group-Asia, Inc (IDG-Asia).The magazine pays to use some of the content of the US *National Geographic Traveler*. *Traveling Scope* (lvyou tiandi), like most Chinese media outlets, has remained a government-affiliated institution, as a subsidiary of a publishing house in Shanghai since it was launched in 1980. These three major players have largely been the trend setters for the development of travel magazines in China over the years (Wang, personal interview, 2003).

The three travel magazines target different segments of readerships. The *Traveler* magazine targets a well-off readership who can 'afford both the magazine and leisure travel' and 'advertising clients who provide high-end commodities and service products' (Guo, personal interview, 2003). It focuses on independent travelers, while *National Geographic Traveler* targets business travelers; *Traveling Scope* serves urban residents, young people in particular. These strategies reflect a trend in the travel magazine market toward segmentation that coincides with increased segmentation in the travel market in general (Wang, personal interview, 2003). In fact, from the mid-1990s, special magazines have emerged for niche travel, such as skiing, golf and adventure trips.

Repositioned cultural geographic magazines are an example of another style of journalistic practice around travel. Amongst them, *Chinese National Geography (zhongguo guojia dili)* is the most popular (Shan, 2002, p. 1).

One feature shared by all these magazines is the intensive use of visuals. Their size has increased from around 45 pages in 1980s to currently 140–170 pages. *Traveler* started to feature foreign destinations when it was launched, and by 2003 almost half of its content was about overseas travel (Wang, personal interview, 2003).

Since the new millennium, new titles have continued to enter the market, targeting more specific groups. For example, *New Voyage* (xin-lvxing), published monthly, targets the high-end business leaders. The *Lonely Planet Traveller China Edition* was launched in 2012 under license from Cite Publishing Ltd, bringing a backpacker and environmental perspective of travel to China. With over 100 titles of travel and travel-related magazines in 2013, competition among travel magazines is increasingly fierce.

Television travel programs

Specialized television programs emerged in the early 1990s as part of a trend toward lifestyle infotainment. China Central Television Station (CCTV), the national broadcaster, was the first to create travel programs, followed by regional and local television stations. Apart from domestic programs, Chinese audiences in some parts of the country are now also able to watch a selection of television travel programs on foreign services such as the Discovery Channel, which have been authorized to broadcast in parts of China since 2002.

Over the years, formats and content of television travel programs have evolved, and travel programs have shifted from trade promotion of tourist sites, attractions and tour packages to more sophisticated and popular infotainment. CCTV, the regional Zhejiang Satellite Television Station and China Travel Television Station are examples for these changes.

For example, Chia Tai Variety Show, a 50-minute travel quiz show launched in 1991 on CCTV, is the first and most popular television travel program in China. For the first few years, the program used travel documentaries imported from Taiwan and Japan. From the mid-1990s, the program started to produce documentaries of its own, first on domestic destinations and later on overseas destinations. According to CCTV's official website, the program had featured various destinations in 120 countries and regions by 2010.

One of its segments, 'A Wonderful World', is a trademark of the show and 'has opened a window to the world outside, especially during the early 1990s when only a few had chances to travel overseas' (Wang, 2003, p. 4). The segment changed its title to 'My Wonderful World', as the program started to solicit home videos of personal travel experience from a nationwide audience from 2001, as a response to increasingly popular outbound travel by the Chinese people.

Meanwhile, provincial television stations started to make travel programs of their own, and many broadcast their programs nationwide through satellite transmission.[2] Between 2002 and 2009, more than 20 provincial stations televised their travel programs nationwide, and among them adventure-seeking programs using Western formats like that of the reality show Survivor became popular.

From 2002 to 2006, Zhejiang Satellite Television aired 'Travel Edition'. Broadcasting for a total of nine hours each Saturday, the program had six major segments, comprised of studio talk shows, documentaries, live reporting, personal travel accounts and entertainment games. It provided extensive travel information, gave concrete travel advice and discussed controversial topics concerned with travel and tourism (Wei, 2009).

Hainan Satellite Television went a step further. Relying on the abundant tourism resources of Hainan as a holiday sea resort, the provincial station turned itself into a specialized travel channel. In July 2002, Hainan Satellite Television became China Travel Satellite Television (lvyou weishi), jointly owned by the Hainan Television Station and China Television Media Co. Ltd. It is the only channel exclusively dedicated to travel, leisure and entertainment in China. All its programs are produced in Beijing and transmitted via Hainan TV's satellite channel. Focusing on travel, fashion, leisure and entertainment, the satellite broadcaster targets an urban audience aged between 20 and 50 who can afford travel and other leisure activities (Jia, 2002).

The launch of this specialized travel service was expected to provide a model for the Chinese television industry to respond to the niche consumer markets, according to Hu Suo, a television travel program producer with northwestern Guizhou Satellite Television (Hu, personal interview, 2003). With investment from Poly Group, a leading private investor in cultural industries in China, since 2004, the travel channel has included more entertainment programs in its broadcasting.

Yet, the popularity of television travel programs has declined over the past few years. One reason is competition from the Internet (Wu, 2007). Some producers also lament a lack of innovation in formats and content connected to a shortage of experienced travel program producers, editors and reporters (Wei, 2009; Zhuan, 2010). Audiences in some part of the country may also switch to programs on foreign services such as the Discovery Channel.

Travel journalism on the Internet

The development of travel journalism in the traditional media has been accompanied by online travel journalism since the late 1990s. By 1999, most major news media institutions including news agencies, newspapers, radio and television stations had an online presence. Many of them included travel as a major category of their online content. Commercial news portal websites like sina.com and sohu.com had specified travel categories as well. Specialized travel websites of different background, like Shanghai-based Ctrip Travel Web and the Guangzhou-based Yahtour, entered the travel information market, providing both travel information and services.

Traditional media were able to expand their publishing offers online. The websites launched by travel magazines are by no means a replica of the magazines themselves. The website of the *Traveler* magazine, for instance, provides news, discussion forums, a photo gallery and even streamed video, apart from the content of the print magazine. In addition, the Internet allows non-journalistic institutions, like tourism authorities, tour operators, airlines and hotels to publish information directly. Moreover, with the help of the Internet, individuals are able to do self-publishing. One travel blog, 'Sun Road', for instance, which was created by a young professional in Shenzhen, southern China, in 2002, focused on the travel experiences of its creator and his thoughts about travel and life. Since then, more individual travel blogs have sprung up, and some of them have turned into travel websites, as in the case of Mafengwo, a travel website in China that was in the top 10 in terms of visitor numbers in January 2013; the website is well known for its user-generated content (Xin, 2013).

On their travel websites, news media institutions, travel media outlets, commercial news portals and travel operators open forums and chat rooms for people to paste stories of their travel experiences, offer travel tips and look for travel companions. Traditional media institutions like

People's Daily and CCTV are now using Weibo, a Chinese version of Twitter, as a means of engaging their audiences.

By enabling a new and participatory platform for the expression and exchange of opinions, the Internet has brought profound changes to travel journalism and the media in China at large. Traditional media no longer monopolize discourses on travel and tourism, but users participate in media discourses, as citizen journalists, contributing to 'replacement of journalism as the dominant public discourse and an arbiter of the public discourse' (Chalaby, 2000, p. 37). As they do not need to appease the government or please travel industry operators, their accounts are usually more expressive and critical. Their accounts and opinions often seem more convincing compared to those of professional journalists, who may be influenced by public relations requirements of either government or industry. The most frequently visited travel websites in China today are those run by private companies that offer practical advice on value-for-money trips with authentic experiences. The top five include 12306 (http://www.12306.cn), Qunar (meaning 'where to go', http://www.qunar.com), Ctrip (http://www.ctrip.com), Mafengwo (literally meaning 'nestle for ants and bees', http://www.mafengwo.cn) (Experian Hitwise, February 2013). Another travel website, Qyer (meaning 'budget travel', http://www.qyer.com), which provides travel information about 40,000 domestic and overseas destinations, is extremely popular among college students.

Online travel journalism has its limitations, however. Most travel websites only process information from other sources and none of them can be compared to major international travel websites such as Lonely Planet, which can use writers all over the world for latest information. Some rely on user-generated content. These limitations have implications for the accuracy and reliability of information.

Current situation

Increases in the coverage of travel in the news media, the growth of specialized travel-oriented media and the proliferation in form and content of travel reporting and writing indicate that travel journalism in China has become a recognized genre of lifestyle journalism that helps promote personal identities, create 'desires for and "knowledge" of new consumer products and services among aspiring groups, cultivate their "taste" and enhance their skills in "modernising their lifestyles"'(Li, 2012; Xu, 2007).Travel journalism has become a multi-dimensional genre that integrates informative, promotional and investigative and consumer-guide styles of reporting. In terms of writing and reporting styles, new models that combine journalistic values, consumer guide information and entertainment have emerged. This style replaces conventional reporting, which emphasized information and aesthetic values and an understanding of travel as a way of learning (Hu, personal interview, 2003). Travel journalism not only reflects travel development, but also initiates changes by broadening and packaging the concept of travel. By incorporating new experiences of leisure, travel and consumption, it influences choices of travel consumption.

In this process of development, travel journalism has absorbed global influences in both form and content, merging with international services to present an ever-expanding range of experiences to different consumers. Travel media, like other lifestyle media in China, are catching up with the West, where lifestyle content has boomed since the 1950s (Chalaby, 2000). Travel magazines in China have international outlooks now. Television travel programs, like other lifestyle programs, are 'adopting formats popular in the West and loading in popular content that is approachable for local viewers' (Xu, 2007, p. 363). Beyond that, travel media are also innovative. Mafengwo, a private travel website mentioned above, has focused on the provision of user-generated content of high quality, a strategy to allure readers and potential writers for the website.

The three-fold expansion of travel journalism

Besides media type, the development of travel journalism was influenced by two other factors: the revenue model of the organizations and the degree of government influence. The media system in China has experienced tremendous changes since the country's opening and economic reforms, particularly after China's shift to a market economy in 1992 and its admission into the World Trade Organization in 2001. These changes led to a three-pronged media system: the 'official' media, as I call them, are state-owned and state-run media institutions, which rely largely on government funding. The 'negotiated' media are a newer class of media which are positioned between government control and the market, as a large number of state media outlets were forced by government edict to commercialize in order to become self-sustaining. The 'flexible' media refers to the latest wave of online media.

Applying this model to travel media, clear distinctions emerge: while the official media maintain an authoritative position for covering areas of industry regulation, and consumer rights protection, the negotiated media offer practical information about new travel experiences.

Meanwhile, flexible media have become a platform where people share personal information and different perspectives on travel. Table 8.1 outlines the interrelations between different types of media, their characteristics, dominant discourses and various functions.

The official media have been historically linked to the Chinese socialist idea of media as an educator and have been used by the Chinese Communist Party and the government to publicize policies and mobilize people (Lee, 2000; Schurmann, 1968). They help maintain government ideologies, contribute to the formation of Chinese national identity and serve national development goals. Official media join the government and the travel industry in creating ideal images of Chinese tourism. They highlight and celebrate Chinese history and culture as well as recent achievements in economic and social development. They represent Chinese national identity through the construction of images of China based on officially approved versions of the past surrounding historical and cultural sites. Some official media institutions that target international visitors such as the official English-language newspaper *China Daily* particularly exemplify this strategy. In its portrayal of Beijing during the SARS epidemic in 2003 as a 'calm' city where 'life goes on', *China Daily* followed the government's guidelines by creating a positive image of the capital before the Olympics. Similarly, Xinhua News Agency and other official media outlets promoted Hainan as a holiday destination, first in 1996 as part of the campaign launched by China National Tourism Administration (CNTA), and more recently in 2009 as an international island resort by the central Chinese government and Hainan provincial government. These official media function as direct promoters of government campaigns. Beyond that, the official media also perform an investigative and watchdog role at times by checking on tourism industry operation, advocating consumer rights and regulating tourism development. One example is Xinhua News Agency's investigative report on 1 June 2002, on the excessive demand for travel to the Three Gorges and the subsequent substandard tour operators before the damming of the Yangtze.

Though controlled by the government, official media institutions are not free from market concerns. As their funding from the government is not sufficient to cover all costs, they need to earn extra income to cover operations and salaries. This situation has caused them to produce new formats and content to appeal to wider audiences. Along with other lifestyle content such as fashion, health, homemaking, parenting, personal finance or automobiles, travel content allows for more advertising revenues and other cost-sharing promotional ventures. For example, CCTV

143

Table 8.1 Development of travel journalism in China, 1978–2012

Types of Media	Media Institutions	Communication Directions	Characteristics	Major Discourse	Roles, Functions & Values
Official (since 1978)	Xinhua News Agency, *People's Daily*, CCTV, *Wenhui Daily*, *China Daily*, Zhejiang Satellite TV, etc.	One to many in traditional media platforms	Government-funded, public service work unit subscription & self-produced profits Professional reporting & writing	Government ideology & national identity	Informative, investigative, educational, consumer-guide; for public good
Negotiated (since 1980s)	*Traveler*, *Trends Traveler*, *Chinese National Geographic*, *Trends Magazines*, *Traveling Scope* China Travel Satellite TV, etc.	One to many, seeking new publics	Self-financed, relying on sales & advertising Professional reporting & writing (including freelance)	Modernity (mobility, money, leisure, service, technology); Post-modernity (hybrid forms of tourism, and tourist experience)	Consumer guide, entertaining, educational Private interests + public good
Flexible media (since the late 1990s)	Independent travel websites like Mafengwo, Qunar, Qyer, Ctrip, Yhtour; online forum, Weibo	Peer to peer	Self-funded; interactive communication; user-generated content, non-professional writing	Experiences (modern and post-modern); DIY	Informative; consumer guide, entertaining, individual interest and experience; personal expression

launched the very popular quiz show 'Chia Tai Variety Show' as early as 1991. *People's Daily* and Xinhua News Agency have created specialized travel sections on their websites, as a way to attract more readers. The official media provide consumer guides and entertainment as much as the negotiated media do.

Negotiated media appeared in 1980s, and have developed very quickly since the early 1990s. Media such as specialized travel magazines and newspapers, newspaper travel editions and television travel programs are representatives of this category. They respond to a demand for more specific information about travel and offer consumer guides and entertainment. They provide insights into tourism and travel developments and promote a new consumption-oriented lifestyle. The government control of popular and lifestyle content is not as rigid, as long as these media outlets do not oppose or criticize the government.

The negotiated media mainly rely on circulation and advertising, impelling them to maintain good relations with their advertising clients. It is always an issue for the negotiated media to balance audience tastes and advertisers' interests. Several media executives mentioned that travel media need to cooperate and, quite often, compromise with the travel industry operators or include content favored by their advertising clients. This practice has challenged the editorial independence of travel media. Nevertheless, audience interest remains a prime concern, as they need to maintain circulation and ratings for advertising revenues.

The flexible media represent the latest stage of media development in China. They emerged in the late 1990s – soon after the beginning of the personal use of the Internet in 1996 – and have since gained increasing popularity, especially among the young and educated. They have contributed significantly to the expansion of travel journalism in China, not only in size but also in form and content.

Online media are flexible in the sense that they offer space for personal expressions and alternative voices, which traditional media fail to provide due to their more direct control by the government. However, online media are not free from government control either. Tension and debate have long existed around free expression of opinions. While many users welcome online opinions as alternatives to the official voices, the government sees especially criticisms of the party and the government or advocacy for democratization, as a threat to its leadership. The dual strategy of censorship and development – filtering and surveillance of alternative voices on the one hand and encouraging the Internet use among its people on the other – is manifestation of the government's ultimate goal to use the Internet to 'reap major economic and political benefits' (Shie, 2005, n.p.). In fact, the government is 'successfully harnessing information technology to maintain its political monopoly' (Shie, 2005, n.p.). Its control of the Internet through myriad regulations, places restraints on all involved: from Internet content and service providers, cybercafés and website creators, to the individual subscriber and user. This policy has considerably reduced the space for alternative opinions. The establishment of the State Internet Information Office in May 2011, which has the same status as the State Information Office under the Chinese State Council, is further evidence of the government's control of the Internet. It directs and coordinates online content management and approves of business related to online reporting (*China Daily*, 4 May 2011).

Yet, despite government control, online communication continues to grow, as it is also necessary for the government to leave some space for personal expressions and alternative voices as an early gauge for possible public discontent. As observed by many critics, the government successfully harnesses information technology to maintain its political monopoly, while the Internet contributes to China's political transformation (Chen, 1999; MacKinnon, 2012; Shie, 2005). It remains a positive force for economic development, improved quality of life and better governance (Shie, 2005). Compared with the print and broadcast media in China, online media therefore enjoy a certain degree of flexibility and freedom. This is particularly the case with popular and lifestyle journalism including travel journalism, as the government mainly does not impose restrictions on popular content online, except for crackdowns on pornography, violence and fraud.

Overall, travel media have expanded over the years in all types of media: 'official', 'negotiated' and 'flexible' media. The boundaries between these types of media are not always clear-cut. While government influence is one limiting aspect, increasing commercialization also challenges the independence of many travel media. For instance, behind the *Trends* Group magazines' Australian holiday promotion in 2003 was a complex array of networks and relationships with different parts of the travel industry.

In many cases, negotiation with the market has resulted in a gradual blurring of the boundary between editorial and advertising content, as evident in the travel sections of the *Yangcheng Evening News* and the *Guangzhou Daily* (Li and Xiao, 2003). It is increasingly difficult for travel journalists to retain an independent position as they are enmeshed in public relations demands of commercial travel operators. This opens up important questions about ethical practices of travel journalism as well as journalism at large.

Travel journalism and national identity

Tourism itself is part of national identity formation and promotion. For instance, as far as inbound tourism is concerned, the government, the travel industry and the media work together to brand, promote and sell China to the international travel market. Similarly, domestic tourism development provides residents with opportunities to experience the nation's natural and cultural heritage and strengthen patriotic connections. Foreign travel can draw attention to the relationship between national identity of individuals and the image of China as a whole. Similarly, reporting about tourism and travel can lead to discussions on national identity and the image of a country.

In the media coverage on travel, the issue of national identity was a common topic. It was explicit, for instance, in *Beijing Evening News'* treatment of the conduct of Chinese tourists groups traveling overseas (February 2003). At one level, the discussion of the manners of Chinese tourists, a series of articles by editors along with summaries of readers' input, was about details, such as how to negotiate a pedestrian crossing, how to dress properly or avoid speaking loudly in public spaces. At another level, this coverage negotiated Chinese identity in relation to others. While the state used to dominate the formation of Chinese national identity, here the travel industry, the media and individual consumers worked with the state ideology to produce a sense of self and national identity. For example, the *Beijing Evening News* discussion of travel etiquette revealed a strong anxiety among readers over the image of the Chinese nation in the world and whether the country can be considered 'advanced' or not. Travel journalism became a platform for discussion of national identity and Chinese citizenship, where editors, reporters and readers interacted with one another through telephone calls, emails and faxes. Citizenship became 'enacted', of interest since citizenship as a concept comes with a distinctive meaning in China. In contrast with the Western-liberal notion of active citizenship, in China, the term often implies fitting in with social norms (Ma, 1996). To cultivate Chinese citizenship, as many of the participants in the discussion observed, it was deemed necessary to strengthen education and improve cultural and ethical standards in society. Travel journalism took on the role of an educator, teaching people how to conduct themselves both as 'qualified' citizens in their home country and as global citizens while traveling abroad.

Travel journalism can also serve as a platform where patriotism and even nationalism find an expression. An example here is the media coverage of the public outrage about the massive sex scandal involving a group of over 400 Japanese male tourists who came to prostitute Chinese girls in Zhuhai, south China in September 2003 (*China Youth Daily*, 28 September 2003). The Japanese were accused of insulting Chinese people at a sensitive time. The conduct by the Japanese tourists was seen by the public and in the media as imperialism in both an economic and a cultural sense. The coverage was anchored in a historic discourse tied to the Chinese resistance against the Japanese invasion between 1937 and 1945.

In general, travel journalism conveyed a mix of national pride and admiration of the Western developed countries. For instance, a strong sense of national pride prevailed in the coverage of surging domestic travel consumption and growing popularity of overseas travel among Chinese people in the general news media. The fact that more than 100 countries are now open to Chinese tourist groups is taken as evidence of China's increasingly high profile and acceptance in the international community.

However, national pride can also make way for an ardent admiration of the West and a deep awareness of China's lack of development. An example was the coverage of Australia in all types of Chinese media since Australia was first opened to Chinese tourist groups as a so-called Approved Destination Status country in 1999. Although it is typical for travel journalism to highlight and celebrate the Other (Fürsich, 2002), the way Chinese travel journalists represent developed countries is different from other foreign destinations. The distinction mainly lies in the uncritical romanticizing of developed countries. Many portrayals of developed countries are full of praise for the affluence, prosperity, the clean environment, good quality of life and the presumably high cultural and ethical standards of locals. Descriptions usually lead to direct comparisons between China and developed countries in these aspects.

Such mixed feelings revealed in travel journalism are closely related to a specific historic re-telling of the rise and fall of the Chinese empire in the past. They are reflections of a narrative about another journey – the enduring journey of the Chinese nation (Zi, 2011) – that is also being told in Chinese travel journalism. None of these sentiments are new in current writings about travel to Western countries. They can be traced back to travel writings and accounts by those Chinese, students, men of letters, government envoys alike, who had the chance to visit and inspect Western countries during the late Qing Dynasty (1616–1911). Their travel reports depict changing attitudes towards Western modernity, from surprise and suspicion to acceptance, along with an increasingly introspective scrutiny of their own country. Such a reflection can still be seen in travel accounts by Chinese writers visiting Western countries in the 1980s (Qiao, 1994). Thus, resurfacing in the contemporary discourse of travel media it is an enduring rather than emergent representation (Good, 2013, p. 297).

Conclusion: travel journalism as a site for studying transformations in contemporary China

Development of travel journalism in China has taken its own path over the past three decades. It has established itself as a recognized genre of popular and lifestyle journalism due to the rapid tourism growth along with market-driven media reforms in China. Travel journalism has experienced a three-fold development within the official, the negotiated and the flexible media. These divergent transformations have allowed for a wide range of information, advice and discussion. In the case of the official media, the information has been framed by concerns to regulate; negotiated media have had more room for commercial promotions, while the flexible online media have allowed non-professional participation. Thus, the development of travel journalism in general reflects the evolution of the Chinese media system from a propaganda institution to a complex media industry, and, more recently, to a platform for personal expression and alternative voices. The evolution of travel journalism echoes a changing relationship between the state, the media and consumers since the economic reforms, especially since China announced its shift to a market economy in 1992.

Travel journalism has not only reflected developments in the tourism industry, but has also initiated changes by defining, broadening and modifying the concept of travel. It has incorporated new experiences of leisure, influenced choices of travel consumption, and promoted transformations towards a consumer society and commodified lifestyles.

A strong impetus for the growth of travel journalism has been the government support for the development of the tourism market. Accordingly, discourses of Chinese modernization are carried through to the popularization of travel as a subject in the media. Moreover, Chinese travel journalism offers advice on social conduct for travelers, both in domestic and international situations, and it influences national self-perceptions and international outlook. In this way, travel journalism provides a platform for the public discussion of national identity and belonging in relation to the world at large.

Travel media also have their blind spots. While journalists link the increasing popularity of leisure travel and holiday-making to the growing affluence and prosperity of the Chinese society, their coverage also reveals the growing regional disparity between urban and coastal areas of China and its rural and inland provinces, and the widening gap between the rich and the poor as China shifts from a planned to a market economy. Media coverage of travel since the mid-1990s has focused on urban travel consumption, especially about outbound travel from major Chinese cities and coastal provinces. This does not mean that the media has neglected travel consumption in rural and inland areas. Rather, the lack of coverage of travel consumption in rural and inland areas reflects the uneven economic development in different regions of the country. Developing in a broader context of social, economic and cultural changes, travel journalism provides a valuable gauge for the study of the transformations in Chinese society and Chinese lifestyles. In a time of dramatic societal transformations, travel media consistently negotiate and co-create an idea of what it means to be a tourist, a consumer and ultimately what it means to be Chinese.

Notes

1. The chapter is based on the author's PhD dissertation, completed in 2005, with updated information. The dissertation combined case studies, discourse analyses and in-depth interviewing.
2. Provincial television stations usually broadcast to local audiences. Each province is allowed to have a selected set of programs broadcast nationwide, and program selection is contingent upon a collective agreement reached among provincial stations to take these programs.

References

Bao, Jiannu (2005) *Going with the Flow: Chinese Travel Journalism in Change.* Unpublished dissertation. Brisbane, Queensland University of Technology.

Chalaby, Jean K. (2000) 'Journalism Studies in an Era of Transition in Public Communications', *Journalism: Theory, Practice and Criticism*, 1.1, pp. 33–9.

Chen, Yan (1999) *Internet Changes China*, Beijing: Peking University Press.

Chen, Zhen (2013) 'The Miscellany of the *Traveler* Magazine', *Traveler*, 205.1, p. 1.

China Daily (2011) *China Sets Up State Internet Information Office.* 4 May 2011. Available at: http://www.chinadaily.com.cn/china/2011-05/04/content_ 12440782.htm

China Youth Daily (2003) 'Japanese Tourist Group Comes on Zhuhai Trip to Prostitute Chinese Girls on Humiliation Day', 28 September, p. 1.

Experian Hitwise (2013) *Visitor Numbers of Domestic Travel Websites in January 2013*. Available at: http://www.experian.com/hitwise

Fürsich, Elfriede (2002) 'Packaging Culture: The potential and limitations of travel programs on global television', *Communication Quarterly*, 50.2, pp. 204–26.

Good, Katie Day (2013) 'Why We Travel: picturing global mobility in user-generated travel journalism', *Media, Culture & Society*, 35.3, pp. 295–313.

Guo, Ziying (2003) Personal interview, 16 January 2003, Beijing.

Hu, Peijiong (2003) Personal Interview, 23 January 2003, Shanghai.

Hu, Suo (2003) Telephone Interview, 25 February 2003, Beijing.

Jia, Yi (2002) *Doing the Best: China Travel Satellite Television*. Available at: http://www.qianlong.net.com/66/2002-10-11/33@462772.htm.

Lee, C-C (2000) 'China's Journalism: the emancipatory potential of social theory', *Journalism Studies*, 1.4, pp. 559–75.

Li, Shuang (2012) 'A New Generation of Lifestyle Magazine Journalism in China: The professional approach', *Journalism Practice*, 6.1, pp. 122–37.

Li, Xiaodong (2003) Personal Interview, 16 March 2003, Guangzhou.

Li, Xiaodong and Xiao, Wei (2003) A Comparative Study of Editorial Styles and Reception of *Yangcheng Evening News* and *Guangzhou Daily*. Available at: http://www.chinaadren.com/html/file/2005-3-4%5C2005342347185102.htlm

Ma, Changshan (1996) 'Civic Consciousness, the Driving Force of Legal Construction in China', *CASS Journal of Law*, 18.6, pp. 3–12.

MacKinnon, Rebecca (2012) *Consent of the Networked: The Worldwide Struggle for Internet Freedom*. New York: Basic Books.

People's Daily (1979) 'Great Potential in Tourism Industry Development', 19 February, p. 1.

Qiao, Chunlei (1994) 'Image of the West in Writings about Travel Overseas by Chinese in 1980s', *Journal of Tianzhong*, 21.6, pp. 82–4.

Richter, Linda K. (1989) *The Politics of Tourism in Asia*, Honolulu: University of Hawaii Press.

Schurmann, Franz (1968) *Ideology and Organization in Communist China*. Berkeley, University of California Press.

Shan, Zhiqiang (2002) 'Fifty Years of Chinese National Geography', *Chinese National Geography*, 12, pp. 1–2.

Shie, Tamara Renee (2005) *Beijing Ahead in the Internet Game*. Asian Times Online, 31 August, 2005. Available at: ttp://www.atimes.com/atimes/China/GH31Ad01.html

Wang, Yan (2003) Personal Interview, 16 January 2003, Beijing.

Wei, Chengyuan (2009) 'Problems with Television Travel Programs and Possible Solutions', *Young Journalists*, 8, pp. 59–60.

Wu, Ning (2007) 'Change of Information Forms and Tourism Marketing', *Journal of Tourism Tribune*, 22.4, pp. 6–7.

Xin, Jianjun (2013) 'Making the Travel Website Mafengwo for Fun', Business Value, March 12, 2013. Available at: http://content.businessvalue.com.cn/post/9611.html.

Xu, Janice Hua (2007) 'Brand-new Lifestyle: Consumer-oriented Programs on Chinese Television', *Media, Culture and Society*, 29.3, pp. 363–76.

Zhuan Jiaoai (2010) Problems with Television Travel Programs and Possible Solutions, Netease Blog. Available at http://tatabu1988.blog.163.com/blog/static/65869130201061953017167/.

Zi, Zhongyun (2011) *View the World at Ease - Self-selected works of Zi Zhongyun*, Guilin: Guangxi University Press.

Part III
Destination Unknown: Content and Representations

9
Along Similar Lines: Does Travel Content Follow Foreign News Flows?

Folker Hanusch

Introduction

One of the key aspects of travel journalism that has attracted some significant amount of scholarly attention relates to the field's role in mediating distant cultures. This may be expected, as one of the key roles of travel journalism is to represented 'otherness', to explore other countries and cultures and to bring them home to audiences. If tourism's purpose is to escape everyday existence, travel journalism's role is to tell people about the variety of ways in which they can do that. Most approaches in this strand of travel journalism studies hail from the cultural studies tradition, have been conducted by tourism scholars and focused predominantly on the content of travel journalism, concerned with the end product and its potential effects on audiences.

Journalism and communication researchers have, by comparison, somewhat neglected this line of inquiry, despite the fact that there exists a long line of research into how other countries are represented in hard news reporting, namely foreign news flow research. Foreign news flow studies can trace their origins to the 1920s and 1930s, with a particularly strong interest leading to a multitude of studies in the 1970s and 1980s. These studies were concerned with what many perceived as an imbalance in the flow of foreign news. Countries in the developing world were systematically disadvantaged, receiving a comparatively small amount of coverage; and what little coverage there was, was typically bad news. Over the past few decades, however, the contents of news media – particularly in developed countries – have changed considerably, due to the phenomenal rise of lifestyle journalism, which includes travel journalism. This rise has come at the same time as a decline in foreign news reporting, brought on by cost-saving measures as traditional media have struggled to remain economically viable. Thus, as one American travel editor has argued, 'in this day of disappearing foreign bureaus, the travel section is many papers' only in-house window on the world at large' (Swick, 2001, p. 65). Just how such travel sections cover the world, however, has not been the subject of sustained research efforts.

This chapter therefore aims to explain how combining tourism studies approaches and research on foreign news flows can advance our understanding of travel journalism as a field. It reports on the results of a major comparative study into the visibility of what I call the 'travel journalism geography' in newspaper travel sections in Australia, Britain, Canada and New Zealand. I will demonstrate that many of the inequalities that have existed in foreign news flows also extend to travel journalism. In this way, one may argue that travel journalism is missing out on an opportunity to provide a more balanced picture of the world.

Travel journalism and foreign destinations

While travel journalism research is still in its early phase of development, recent years have seen an increase in the literature on the topic, with the analysis of travel content the most commonly applied method (Hanusch, 2010). Only some of this work has occurred from a communication or journalism studies perspective, however. Most studies have been conducted by tourism scholars concerned with whether travel journalism content was in line with marketing expectations. For example, Pan and Ryan (2007) have developed an analytical framework that would allow destination marketers to assess travel stories' value to them. Such studies could include, for example, the ways in which travel stories associate particular sensory allusions with certain destinations (Pan and Ryan, 2009). Travel writing has also been argued to aid in educating tourism students and add to an understanding of other cultures (Armstrong, 2004). In addition, the portrayal of foreign destinations through travel writing has been of concern to critical tourism scholars. This work has followed two main lines of enquiry. First, research has been interested in the extent to which certain countries are covered in general; and second, other research provided a deeper reading of how host cultures are portrayed within reports.

The first strand of studies has demonstrated that travel journalism tends to focus on a relatively narrow set of destinations. For example, a small study of travel segments broadcast on the Travel Channel in the US found that 50 per cent of them focused on the US, and a further

30 per cent focused on Western Europe (Mahmood, 2005). A study of three Australian newspapers' travel sections showed that North America, Europe and Southeast Asia were the most regularly covered regions (Hill-James, 2006) – showing strong similarities with foreign news coverage in the country. My analysis of six Australian newspapers' travel sections confirmed this, with the three regions accounting for a combined 56.3 per cent of all stories (Hanusch, 2011). Such findings, albeit based on a small set of studies, demonstrate that factors such as a focus on elite nations, cultural proximity and regionalism all appear to be significant criteria for selection in travel journalism as they are in foreign news reporting.

The second strand of inquiry in this field has shown that host cultures are often marginalized, with a number of studies finding that locals only tend to be included when they are involved in the tourism industry. Typically travel journalism focuses on the experience of the presenter or author rather than host cultures (Dunn, 2005; Hanefors and Mossberg, 2002; Santos, 2004, 2006). Galasinski and Jaworski's (2003, p. 131) analysis of travel stories in the *Guardian* found that citizens of tourist destinations were typically represented as homogeneous groups, representatives of (stereotypical) national or community characteristics, or as '"featureless" helpers to the travelers'. My study of Australian newspaper travel stories found that three out of every five stories included no quotations, and the vast majority of those that did, only quoted locals working in the tourism industry (Hanusch, 2011).

This chapter is concerned primarily with extending the first aspect, that is evaluating the amount of reporting on various destinations. As opposed to earlier studies, it anchors the analysis within research on foreign news flows, which has a long tradition in mass communication research.

Foreign news flows

The study of foreign news can be traced back more than 80 years, when scholars became interested in the image that newspapers presented of the world around us (Woodward, 1930). It only reached large global popularity during the 1960s and 1970s, however, when developing countries called for a New World Information and Communication Order (NWICO), which culminated in the MacBride report of UNESCO in 1980. Many from the so-called Third World believed they were portrayed unfairly in the foreign news of Western countries. They argued that their countries were rarely covered, and that, when they were, the focus was on wars, crises and disasters (Hachten, 1999). NWICO advocates' ultimate aims were for developing countries to produce more stories of their own and create a more positive image of themselves. There was strong opposition to the proposal from Western countries, who saw it as an attempt to politicize foreign news and sacrifice press freedom. The debates reached no real resolution and eventually Britain and the US left UNESCO (Hachten, 1999).

The debate over the NWICO motivated a generation of researchers to examine in some considerable depth the flow of foreign news. Between 1970 and 1986 alone, at least 150 research papers were published on the topic, including two worldwide studies under the auspices of the International Association for Mass Communication Research (IAMCR) and UNESCO (Tsang et al., 1988). In fact, at the time Wilke (1987, p. 147) argued that 'no subject in communication research in recent years has stimulated greater interest on a world-wide level than questions of foreign news reporting and international news flow'.

The first UNESCO study, conducted in 1979 across 29 countries, found that, primarily, foreign news coverage concentrated on events in a country's immediate geographic region (Sreberny-Mohammadi et al., 1984). Major political powers were the second most frequent topic, followed by countries that experienced crises or disasters. Countries on the periphery, which were of little importance in the global political system, rarely received coverage, unless they suffered a catastrophe. The researchers also found an imbalance of news flow, with developing countries receiving far more news about developed countries than vice-versa. In the second major international study, conducted in 1995 and including 46 countries, researchers found geographic proximity and national linkages remained the dominant news values in media worlds mostly defined by politics and economics (Stevenson, 1997). Further analysis by Wu (2000, 2003) concluded that trade, population, the presence of international news agencies as well as geographic proximity were principal determinants of whether a country would make the news. Similarly, Ito's (2009) analysis of the data from the 1995 study found five factors influenced foreign news flows: the existence of an international news agency in the covered country, the amount of trade and geographic distance between countries, but also a common language and the covered country's defense budget (ostensibly a sign of military power). A meta-analysis of 55 research papers showed that factors influencing foreign news selection included the gross national product of a nation, trade volume, regionalism, population as well as geographic size, geographic, political, economic and cultural proximity, eliteness, communication resources and infrastructure (Wu, 1998).

As researchers moved beyond studies of the print media and into television and, later, online news, they came to similar conclusions, noting that even new media did not drastically change the long-established trends. Wilke et al.'s (2012) recent study of 17 countries' television news confirmed that regionalism, the role of superpowers and crises still played a major role in news decision-making. Overall, Europe was the most heavily covered continent, with North America and the Middle East in second and third place. In terms of individual countries, the US and Britain were the most frequently covered countries, leading Wilke et al. (2012, p. 319) to conclude that 'the picture of the world is still uneven, particularly with respect to regions of the world that are totally underrepresented'. Online, these traditional imbalances are also still apparent (D'Haenens, Jankowski and Heuvelman, 2004; Gasher and Gabriele, 2004; Himelboim, Chang and McCreery, 2010; Wu, 2007). For example, Wu (2007) found that trade volume and the presence of international news agencies were still important predictors of coverage in the online environment.

Yet, the vast majority of these studies of foreign news flows have focused only on the news sections of various media. At a time when space for foreign news has become smaller (Riffe et al., 1994), and other types of journalism such as lifestyle journalism are taking up an increasing amount of space across all news media, it is imperative that sections such as travel are examined in more detail in order to determine their contribution to our mediated ideas of the world.

This study was guided by three main research questions. The first question asked which world regions and countries newspaper travel stories actually focused on. The second question was concerned with determining whether there were any similarities or differences across different countries in terms of their geography of travel journalism. Finally, the third research question asked how the geography of travel journalism compares with what we know about foreign news geography.

Methodology

In order to examine these research questions, travel sections from two newspapers each from Australia, Britain, Canada and New Zealand were selected during the first half of 2009. The countries were chosen purposefully in line with a 'most-similar systems design' (Przeworski and Teune, 1970, p. 32), which is based on 'the belief that systems as similar as possible with respect to as many features as possible constitute the optimal samples for comparative inquiry'. This approach allows researchers to minimize the number of contributing variables by choosing countries that share a number of characteristics. As Australia, Britain, Canada and New Zealand all share similar political and cultural characteristics, have similar media systems and historical connections, they were deemed an appropriate sample for further examination.

Within each country, the two leading providers of quality newspaper travel stories were chosen. Quality newspapers were chosen because of their influence as agenda-setting media, as well as the fact there are no real tabloid newspapers in New Zealand, which could have led to an uneven analysis had tabloid newspapers from other countries been included. The chosen newspapers are *The Australian, Sydney Morning Herald* (Australia); *The Guardian, Daily Telegraph* (Britain); *The Globe and Mail, Toronto Star* (Canada); *New Zealand Herald, The Dominion-Post* (New Zealand). The Australian, British and Canadian newspapers were examined on Saturdays, while the New Zealand newspapers publish their travel sections on Tuesdays. The timeframe of analysis was each week from 10 January until 7 July 2009, covering 26 issues of each sampled newspaper. Thus, a total of 208 travel sections were examined for analysis.

Only travel stories that focused on an individual country or domestic region were included in the study. Stories that focused on a theme, and discussed a variety of countries or domestic regions, were excluded. Editorials, reader questions, celebrity travel, Q&As or listings of travel tips and deals were also excluded. Stories were further coded in terms of whether they covered domestic or foreign travel destinations, and, if the latter applied, which country the destination was in. Only one country could be coded per story, any multiple-country stories, as discussed earlier, were excluded from the study. Stories were coded by three coders and inter-coder reliability tests, conducted on 10 per cent of the overall sample, resulted in a Krippendorff's alpha score of 0.976 for country of location and 0.995 for foreign/domestic news.

A total of 1074 travel stories across 208 issues were examined for analysis across the eight newspapers, equating to an overall average of 5.2 stories per newspaper per week. Almost half of the worlds' countries (116) were covered during the research timeframe. Australian newspapers had the most dispersed coverage, reporting on 71 different countries, followed by Britain and Canada who each covered 59 different countries each, and New Zealand, which only reported from 42 different countries. The split between domestic and foreign travel stories differed across the countries. Canada had the most international outlook, with 83.2 per cent of stories about another country. It should be noted, however, that a large percentage of those were about the nearby US. On the other end of the spectrum, Australian newspapers had the smallest proportion of foreign travel stories at only 65 per cent. Britain (73.5 per cent) and New Zealand (69.7 per cent) took the middle position.

In the following, the analysis of the results is divided into two sections. First, I examine which regions were covered the most frequently across the four countries. In the second part, the prominence of specific countries is explored.

Most frequently covered regions

In terms of world regions, Table 9.1 shows some important similarities in travel journalism geography to what we know from foreign news flows.

Europe is the most heavily covered continent, followed by the Americas, Asia, Oceania and Africa. Much like in foreign news, Africa continues to remain a blank spot even in this softer form of journalism. Newspapers in Australia, Canada and New Zealand accorded the continent only a maximum of five per cent of all stories, while in Britain it fared slightly better, receiving 10 per cent of coverage. Looking at the results only through the five major regions hides a more nuanced picture, however. When we examine sub-regions, we can find a clear preference for Northern and Western Europe, Northern America and Southern Europe. None of the other sub-regions received more than 10 per cent of coverage overall. This result is somewhat in line with official tourism statistics, which show that Europe is the most popular tourism region in the world (World Tourism Organization, 2012).

In looking at individual countries' reporting of the world, regionalism – found to be a strong determinant in foreign news reporting – also appears to be important in travel journalism. Each of the countries examined here tended to focus on a region close to them, which is not surprising given proximity makes it easier for readers to travel there. Australian newspapers focused more than one-third of their coverage on Asia (37.1 per cent of all stories), while New Zealand newspapers gave a similar amount of prominence to Oceania. Canadian newspapers concentrated almost half of their stories on the Americas, and British newspapers a similar amount on Europe.

At the same time, in terms of sub-regions, we can also see that in Australia, Northern and Western Europe received the same amount of

Table 9.1 Coverage of world regions by country (in per cent)

	Australia	Canada	New Zealand	UK	Total
Africa	**5**	**3.1**	**4.3**	**10**	**5.7**
Eastern Africa	1.3	1	0.7	3.5	1.7
Middle and Western Africa	0	1.6	0	1	0.6
Northern Africa	2.5	0	0.7	4	1.9
Southern Africa	1.3	0.5	2.9	1.5	1.4
Americas	**16.7**	**43.5**	**17.9**	**25.5**	**25.9**
Northern America	9.6	27.5	14.3	14	16
Central America and Caribbean	2.5	8.8	0.7	5.5	4.6
South America	4.6	7.3	2.9	6	5.3
Asia	**37.1**	**21.8**	**17.9**	**12**	**23.3**
Central and Western Asia	2.9	8.3	1.4	4.5	4.4
Eastern Asia	10	5.7	6.4	1	6
Southern Asia	7.1	3.1	5.7	4	5
South-Eastern Asia	17.1	4.7	4.3	2.5	7.9

163

Europe	**29.2**	**28.5**	**25**	**45.5**	**32.5**
Eastern Europe	3.3	1.0	1.4	1.5	1.9
Northern and Western Europe	17.1	16.6	12.9	26.0	18.5
Southern Europe	8.8	10.9	10.7	18.0	12.0
Oceania	**12.1**	**3.1**	**35.0**	**7.0**	**12.7**
Australia and New Zealand	5.4	3.1	26.4	6.5	8.9
Pacific Islands	6.7	0.0	8.6	0.5	3.7

coverage as South-Eastern Asia, and Northern America followed closely behind East Asia. Other regions – including the proximate and popular holiday destinations in the Pacific Islands – received very little coverage. In Canada, the US dominated coverage at more than one quarter of all stories – more than all the Asian countries combined, and almost as much coverage as all of Europe. Somewhat similarly, Australia was the focus for New Zealand newspapers, making up more than one quarter of all stories. It was followed by Northern America, and Northern, Western and Southern Europe. In Britain, travel sections focused predominantly on their immediate vicinity in Northern, Western and Southern Europe, as well as Northern America. The Asian continent received only 12 per cent of coverage, slightly more than Africa.

Regionalism has frequently been found to be a strong feature of foreign news (Sreberny-Mohammadi et al., 1984; Wu, 1998), and the situation is no different in travel journalism. The only surprise perhaps was the comparatively low attention Australian newspapers gave to New Zealand and the Pacific Islands, arguably Australia's most immediate neighboring region. Nevertheless, Asia has become a region of growing importance for Australia over the past two decades. The results correspond to existing foreign news flow studies in their dominance of stories from Southeast Asia, Western Europe and North America, while Oceania has never been a region of much interest to Australian foreign news (Grundy, 1980; Henningham, 1996; Putnis et al., 2000).

Most frequently covered countries

Once we move beyond the regional analysis to individual countries, an even more nuanced picture of the travel journalism geography emerges. Equivalent to the number one position in foreign news reporting, the US also tops the list of most covered nation in travel journalism, making up 10.5 per cent of all foreign travel stories across the sample (Table 9.2).

This is more than twice the size of coverage of the next country, Australia, which made up 5 per cent of the total. France, the world's leading tourist destination, received 4 per cent of stories, closely followed by Italy. In fifth place was Britain at 3.7 per cent, followed by India and Thailand. One constraining factor here is, however, that the countries whose newspapers were under examination were at a disadvantage, as they could only be covered by three countries as opposed to four, potentially skewing the figures. To account for this, each country's total number of stories received was divided by four, while the totals for Australia, Britain, Canada and New Zealand were divided by

Table 9.2 Top 10 destinations overall

		Number of stories	%
1	US	113	10.5
2	Australia	54	5.0
3	France	43	4.0
4	Italy	42	3.9
5	UK	40	3.7
6	India	27	2.5
7	Thailand	22	2.0
8	Spain	18	1.7
9	Argentina	16	1.5
10	Japan	16	1.5
	Total Top 10	**391**	**36.3**

three. This resulted in a slight shift in rankings. The US still topped the list, followed by Australia. In third place, however, was Britain, which leapfrogged France and Italy. Those two were followed by India and Thailand, with New Zealand now in eighth place, having overtaken Spain and Argentina. This is not a dramatic shift, but it may indicate a slightly more accurate picture of foreign travel coverage.

To detect specific similarities and differences, it is important to examine each country's top destinations in relation to actual travel. Table 9.3 shows a comparison of each country's top 10 destinations as reported through travel stories and by tourist numbers.

The results further cement the trends of the regional analysis. In Australian newspapers, the top two countries were Britain and the US, each of which received 8.3 per cent of coverage. They are followed by Italy, New Zealand and Thailand. Broadly, the coverage appears in line with tourist behavior, but there are also important differences. Britain and the US rank only second and fourth, respectively, in the list of tourist destinations, while they rank first and second in the number of stories. In contrast, New Zealand, by far Australians' favorite tourism destination, ranks only fourth in terms of newspaper coverage. Only two stories were written about Fiji during the research timeframe, despite the fact 3.8 per cent of Australian tourists travel there. Indonesia also received a disproportionately small amount of coverage when compared to tourist numbers.

Table 9.3 Top 10 stories and tourism by country

	Australia					
	Travel stories published			Tourism destinations*		
		Number of stories	%		Departures (in ,000s)	%
1	Britain	20	8.3	1 New Zealand	955.3	16.3
2	US	20	8.3	2 US	500	8.6
3	Italy	14	5.8	3 Indonesia	436	7.5
4	New Zealand	13	5.4	4 Britain	420.2	7.2
5	Thailand	13	5.4	5 Thailand	378.4	6.5
6	India	12	5	6 China	268	4.6
7	Indonesia	11	4.5	7 Fiji	220.9	3.8
8	France	10	4.1	8 Singapore	213.7	3.7
9	Japan	9	3.7	9 Malaysia	205.2	3.5
10	Hong Kong	6	2.5	10 Hong Kong	200.1	3.4
	Total Top 10	128	52.9	Total Top 10	330.8	65
	Overall total	240	100	Overall total	5,843.2	100

Canada

| | Travel stories published | | | Tourism destinations** | |
	Number of stories	%		Departures (in million)	%
1 US	53	27.5	1 US	18	68.7
2 Britain	12	6.2	2 Mexico	1.21	4.6
3 Italy	10	5.2	3 Cuba	0.98	3.7
4 France	7	3.6	4 Dominican Republic	0.88	3.3
5 China	6	3.1	5 Britain	0.87	3.3
6 Israel	6	3.1	6 France	0.74	2.8
7 Turkey	6	3.1	7 Italy	0.36	1.4
8 Argentina	5	2.6	8 Germany	0.31	1.2
9 India	5	2.6	9 China	0.26	1
10 Mexico	5	2.6	10 Netherlands	0.26	1
Total Top 10	115	59.6	Total Top 10	23.87	91.1
Overall total	193	100	Overall total	26.20	100

Table 9.3 continued

		New Zealand				
	Travel stories published			Tourism destinations***		
		Number of stories	%		Departures (in ,000s)	%
1	Australia	37	26.4	1 Australia	944.3	49.2
2	US	15	10.7	2 Fiji	92.3	4.8
3	Britain	8	5.7	3 US	88.1	4.6
4	Italy	6	4.3	4 Britain	87.4	4.6
5	India	5	3.6	5 China	56.3	2.9
6	Canada	4	2.9	6 Cook Islands	55.7	2.9
7	Spain	4	2.9	7 Samoa	41.4	2.2
8	China	3	2.1	8 Thailand	30.6	1.6
9	Cook Islands	3	2.1	9 India	29.0	1.5
10	Fiji	3	2.1	10 Canada	19.4	1
	Total Top 10	88	62.9	Total Top 10	1,444.4	75.3
	Overall total	140	100	Overall total	1,918.32	100

Britain

	Travel stories published				Tourism destinations****	
		Number of stories	%		Departures (in million)	%
1	US	25	12.5	1 Spain	11.58	19.8
2	France	23	11.5	2 France	9.76	16.7
3	Australia	13	6.5	3 Ireland	3.55	6.1
4	Italy	12	6.0	4 US	3.19	5.4
5	Spain	8	4.0	5 Italy	2.61	4.5
6	Germany	7	3.5	6 Germany	2.13	3.6
7	Greece	7	3.5	7 Greece	1.88	3.2
8	Turkey	7	3.5	8 Netherlands	1.84	3.1
9	India	5	2.5	9 Portugal	1.81	3.1
10	Argentina	4	2.0	10 Turkey	1.62	2.8
	Total Top 10	**111**	**55.5**	**Total Top 10**	**40**	**68.2**
	Overall total	**200**	**100**	**Overall total**	**59**	**100**

* *Source*: Australian Bureau of Statistics. Figures express total departures for July 2008 to June 2009. Available at http://www.abs.gov.au/AUSSTATS/abs@.nsf/Previousproducts/3401.0Feature%20Article1Jun%202009?opendocument&tabname=Summary&prodno=3401.0&issue=Jun%202009&num=&view=

** *Source*: Statistics Canada. Figures express total departures for 2009. Available at http://www.statcan.gc.ca/pub/66-201-x/2009000/t044-eng.htm

*** *Source*: Statistics New Zealand. Figures express total departures for 2009. Available at: http://www.stats.govt.nz/browse_for_stats/population/Migration/IntTravelAndMigration_HOTPDec09.aspx

**** *Source*: Office for National Statistics. Figures express total departures for 2009. Available at: http://www.ons.gov.uk/ons/rel/ott/travel-trends/2009/index.html

In Canada, the dominance of the US in travel stories is mirrored by its popularity as a tourist destination. Yet, countries in the Caribbean received disproportionately low coverage. Mexico, Canadians' second-most popular destination, accounted for only 2.6 per cent of travel stories, while Cuba and the Dominican Republic did not even make the top 10 list of travel stories. Only three stories in total were written about the two countries, despite the fact they are highly popular tourist destinations. A similar picture emerged in New Zealand, where Australia was easily the most frequently covered country as well as the top tourism destination. Surprisingly, however, the second-most popular tourist destination Fiji only received 2.1 per cent of all stories. In contrast, Britain and the US, receiving comparable numbers of New Zealand tourists, accounted for a much larger share of stories.

In Britain, seven of the 10 most popular destinations for tourists were also represented in the top 10 list of countries covered in British newspapers; yet once again we see some important differences. The most heavily covered country, the US, received 12.5 per cent of all stories, but only accounted for 5.4 per cent of tourist departures. On the other hand, Spain – which hosts almost one in five British tourists – only received 4 per cent of newspaper travel coverage. Australia, often covered in travel sections, was not even among the top 10 of British tourist destinations, and Ireland, the third most-popular country in terms of departures, did not make the top 10 list of covered countries.

A number of important aspects become apparent in the analysis of the coverage in individual countries. First, it is clear that the US – so dominant in foreign news coverage (Wilke et al., 2012; Wu, 1998) – also is a popular topic of travel stories. More than twice the number of stories overall were devoted to the US, despite the fact it was the most popular tourist destination only in Canada. While the US did rank in the top 4 most popular destinations in each of the studied countries, the fact that other, at least similarly popular destinations, received considerably less coverage suggests that the US receives proportionally more attention in travel journalism as well. Secondly, cultural proximity, an important criterion of foreign news (Galtung and Ruge, 1965) also matters in travel journalism. This is because not just the US, but also Britain received very large amounts of coverage. This can be linked to the fact that the three other countries in this study are all former British colonies who maintain strong cultural, political and economic links with the country. Culturally more distant countries like China, Turkey, Malaysia or Mexico – all four of which rank in the top 10 of the world's tourist destinations – received much less attention. At the same time, not many Canadians, Australians or New Zealanders travel to African countries, and travel sections appear to take the view that they want to offer their readers a service by focusing on destinations to which many travel. 'Exotic' destinations are thus of less relevance overall.

The 'big neighbor syndrome' found in many foreign news flow studies (Hackett, 1989; Hart, 1963; Ichikawa, 1978; Robinson and Sparkes, 1976; Wilke et al., 2012) can also be confirmed in travel journalism. In Canada, the big neighbor US was covered most frequently, accounting for just over one quarter of all foreign travel stories. The situation was similar in New Zealand, where Australia received a high level of coverage. This is a situation very similar to foreign news coverage in Canada and New Zealand (Nnaemeka and Richstad, 1980; Robinson and Sparkes, 1976). At the same time, it needs to be pointed out that the US and Australia are also the most popular tourist destinations for Canadians and New Zealanders, respectively. Thus, the two factors are likely inextricably linked.

Conclusion

The results of this study demonstrate that the geographies of travel journalism and foreign news bear a canny resemblance. Clearly, the factors that appear to apply to foreign news reporting are often transferred to travel journalism. Regionalism, a focus on powerful nations or big neighbors and cultural proximity all play a role in selecting which stories to cover in a travel section. Yet, some factors from foreign news understandably do not apply to travel content. Travel stories tend to shy away particularly from countries experiencing crises or disasters, quite contrary to foreign news. The countries of the Middle East, regularly a key region for foreign news (Wilke et al., 2012; Wu, 1998), did not feature significantly in the travel sections examined here. As one travel journalist once pointed out, the profession's *raison d'être* is that 'travel publications celebrate travel' (Austin, 1999, p. 10) and it is quite clear that crises make a country less appealing as a travel destination.

The latter part of the analysis focused on the role that tourist numbers to a certain country may play in decision-making about travel stories. The results show that tourist numbers can have some influence. For example, Italy was featured in a relatively large number of stories, despite not normally receiving much attention in foreign news. Central America and the Caribbean make up a sizable portion of Canadian travel stories, despite receiving very little attention in Canadian foreign news (Hackett, 1989; Wilke et al., 2012). In Australian foreign news,

Oceania is rarely covered (Dorney, 2000; Putnis, 1998), but just over 10 per cent of overall travel content is focused on the region. While this is still not in line with travel activity, it is at least more frequent than foreign news coverage. However, tourist numbers could not always override other more common factors, such as cultural proximity and political power, which are also often found in foreign news reporting.

What does all this mean for travel journalism's role in an age of declining foreign news coverage? I argued earlier that travel sections potentially had an opportunity to redress some of the imbalances of foreign news reporting. Based on this study, however, it appears that travel journalism is largely replicating these imbalances, thus missing out on an opportunity to perhaps provide a more complex and multi-faceted view of the world. For travel journalists, entertainment and the reporting of foreign cultures are central components of their work (Hanusch, 2010, 2012). If travel media want mainly to entertain readers, they should be able to cover regions or countries that are typically outside the tourists' gaze. But travel journalists also see a crucial need to provide their audiences with information they can use in their travels (Hanusch, 2012), and may thus be compelled to focus on destinations which are already popular with these audiences. The attractiveness of a destination and the amount of coverage does not always appear to be connected, however, and some highly visited destinations – often the less culturally proximate or less powerful ones – are featured less frequently than others.

Perhaps another aspect needs to be taken into account here, too. It may be that an important criterion, beyond the limits of this study, is the budget any particular country has for sponsoring journalist visits to its shores. Travel journalism depends to a large degree on subsidized or free travel (Austin, 1999; Hanusch, 2012), and one could argue that the more powerful and wealthy countries and regions are in a better position to attract coverage, as their tourism boards are more likely to be able to afford comprehensive visiting journalists' programs. Mackellar and Fenton (2000) have noted that the Australian tourism board has a well-funded visiting journalists' program, which may have played a role in Australia gaining substantial attention in the other three countries' travel sections. A further role may be played by individual tour operators, hotels, airlines and other tourism businesses, who often sponsor travel for journalists to visit various countries and region. Certainly future studies would need to take these aspects into account, although it may be difficult to gain access to such data. Such studies could also investigate a wider variety of countries, in order to examine whether the trends in these four relatively similar countries, also apply else-where around the globe. Much like there is room for further sustained research in the travel journalism field more generally, it is clear that the geography of travel journalism also deserves closer investigation. This research will contribute to a more holistic understanding of the way media cover foreign countries and how audiences experience the world through their media.

References

Armstrong, Kate E. (2004) 'Analysis of travel writing enhances student learning in a theoretical tourism subject', *Journal of Teaching in Travel & Tourism*, 4.2, pp. 45–60.

Austin, Elizabeth (1999) 'All expenses paid: Exploring the ethical swamp of travel writing', *The Washington Monthly*, 31.7/8, pp. 8–11.

D'Haenens, Leen, Jankowski, Nicholas and Heuvelman, Ard (2004) 'News in online and print newspapers: Differences in reader consumption and recall', *New Media & Society*, 6.3, pp. 363–82.

Dorney, Sean (2000) 'Where in the world are we!!', *Australian Studies in Journalism*, 9, pp. 15–29.

Dunn, David (2005) '"We are not here to make a film about Italy, we are here to make a film about ME...": British television holiday programmes' representations of the tourist destination', in David Crouch, Rhona Jackson and Felix Thompson (eds) *The media and the tourist imagination: Converging cultures*. London: Routledge, pp. 154–69.

Galasinski, Dariusz and Jaworski, Adam (2003) 'Representations of hosts in travel writing: The Guardian travel section', *Journal of Tourism and Cultural Change*, 1.2, pp. 131-49.

Galtung, Johan and Ruge, Mari Holmboe (1965) 'The structure of foreign news', *Journal of Peace Research*, 2.1, pp. 64–91.

Gasher, Mike and Gabriele, Sandra (2004) 'Increasing circulation? A comparative news-flow study of the Montreal Gazette's hard-copy and on-line editions', *Journalism Studies*, 5.3, pp. 311–23.

Grundy, Bruce (1980) 'Overseas news in the Australian press', in P. Edgar (ed.) *The News in Focus: The Journalism of Exception*. South Melbourne: MacMillan, pp. 120–35.

Hachten, William A. (1999) *The World News Prism: Changing Media of International Communication*, 5th edn, Ames: Iowa State University Press.

Hackett, Robert A. (1989) 'Coups, earthquakes and hostages? Foreign news on Canadian television', *Canadian Journal of Political Science*, 22.4, pp. 809–25.

Hanefors, Monica and Mossberg, Lena (2002) 'TV travel shows – a pre-taste of the destination', *Journal of Vacation Marketing*, 8.3, pp. 235–46.

Hanusch, Folker (2010) 'The dimensions of travel journalism: Exploring new fields for journalism research beyond the news', *Journalism Studies*, 11.1, pp. 68–82.

Hanusch, Folker (2011) 'Representations of foreign places outside the news: An analysis of Australian newspaper travel sections', *Media International Australia*, 138, pp. 21–35.

Hanusch, Folker (2012) 'A profile of Australian travel journalists' professional views and ethical standards', *Journalism*, 13.5, pp. 668–86.

Hart, Jim (1963) 'The flow of news between the US and Canada', *Journalism Quarterly*, 40.1, pp. 70–74.

Henningham, John (1996) 'The shape of daily news: a content analysis of Australia's metropolitan newspapers', *Media International Australia*, 79, pp. 22–34.

Hill-James, Candeeda R. (2006) *Citizen tourist: newspaper travel journalism's responsibility to its audience*. Unpublished PhD Thesis, Brisbane: Queensland University of Technology.

Himelboim, Itai, Chang, Tsan-Kuo and Mccreery, Stephen (2010) 'International network of foreign news coverage: Old global hierarchies in a new online world', *Journalism and Mass Communication Quarterly*, 87.2, pp. 297–314.

Ichikawa, Akira (1978) 'Canadian–US news flow: The continuing asymmetry', *Canadian Journal of Communication*, 5.1, pp. 8–18.

Ito, Youichi (2009) 'What sustains the trade winds? The pattern and determinant factors of international news flows', *Keio Communication Review* 31, pp. 65–87.

Mackellar, Jo and Fenton, Jane (2000) 'Hosting the international travel media: A review of the Australian Tourist Commission's visiting journalist programme', *Journal of Vacation Marketing*, 6.3, pp. 255–64.

Mahmood, Reaz (2005) 'Travelling away from culture: The dominance of consumerism on the Travel Channel', *Global Media Journal*, 4.6. Available at http://lass.calumet.purdue.edu/cca/gmj/sp05/graduatesp05/gmj_sp05_graduateTOC.htm

Nnaemeka, Tony and Richstad, Jim (1980) 'Structured relations and foreign news flow in the Pacific region', Gazette, 26.4, pp. 235–57.

Pan, Steve and Ryan, Chris (2007) 'Analyzing printed media travelogues: Means and purposes with reference to framing destination image', *Tourism, Culture & Communication*, 7.2, pp. 85–97.

Pan, Steve and Ryan, Chris (2009) 'Tourism sense-making: The role of the senses and travel journalism', Journal of Travel and Tourism Marketing, 26.7, pp. 625–39.

Przeworski, Adam and Teune, Henry (1970) *The Logic of Comparative Social Inquiry*, New York: Wiley-Interscience.

Putnis, Peter (1998) 'Australian press coverage of the 1995 Mururoa nuclear test', *Asia Pacific Media Educator*, 5, pp. 38–50.

Putnis, Peter, Penhallurick, John and Bourk, Michael (2000) 'The pattern of international news in Australia's mainstream media', *Australian Journalism Review*, 22.1, pp. 1–19.

Riffe, Daniel, Aust, Charles F., Jones, Ted C., Shoemaker, Barbara and Sundar, Shyam (1994) 'The shrinking foreign newshole of the *New York Times*', *Newspaper Research Journal*, 15.3, pp. 74–88.

Robinson, Gertrude Joch and Sparkes, Vernone M. (1976) 'International news in the Canadian and American press: A comparative news flow study', *Gazette*, 22.4, pp. 203–18.

Santos, Carla Almeida (2004) 'Framing Portugal: Representational dynamics', Annals of Tourism Research, 31.1, pp. 122–38.

Santos, Carla Almeida (2006) 'Cultural politics in contemporary travel writing', *Annals of Tourism Research*, 33.3, pp. 624–44.

Sreberny-Mohammadi, Annabelle, Nordenstreng, Kaarle and Stevenson, Robert L. (1984) 'The world of the news study', *Journal of Communication*, 34.1, pp. 134–38.

Stevenson, Robert L. (1997) 'Remapping the news of the world'. Available at http://www.ibiblio.org/newsflow/results/Newsmap.htm

Swick, Thomas (2001) 'The travel section: Roads not taken', *Columbia Journalism Review*, May/June, pp. 65–67.

Tsang, Kuo-jen, Tsai, Yean and Liu, Scott S. K. (1988) 'Geographic emphases of international news studies', *Journalism Quarterly*, 65.1, pp. 191–94.

Wilke, Jürgen (1987) 'Foreign news coverage and international news flow over three centuries', *Gazette*, 39.3, pp. 147–80.

Wilke, Jürgen, Heimprecht, Christine and Cohen, Akiba (2012) 'The geography of foreign news on television: A comparative study of 17 countries', *International Communication Gazette*, 74.4, pp. 301–22.

Woodward, Julian Lawrence (1930) *Foreign News in American Morning Newspapers: A Study in Public Opinion*, New York: Columbia University Press.

World Tourism Organization (2012) *Tourism Highlights: 2009 Edition*, Madrid: UNWTO.

Wu, Denis H. (1998) 'Investigating the determinants of international news coverage: A meta-analysis', *Gazette*, 60.6, pp. 493–512.

Wu, Denis H. (2000) 'Systemic determinants of international news coverage: A comparison of 38 countries', *Journal of Communication*, 50.2, pp. 110–30.

Wu, Denis H. (2003) 'Homogeneity around the world? Comparing the systemic determinants of international news flow between developed and developing countries', *Gazette*, 65.1, pp. 9–24.

Wu, Denis H. (2007) 'A brave new world for international news? Exploring the determinants of the coverage of foreign nations on US websites', *International Communication Gazette*, 69.6, pp. 539–51.

10

'Out There': Travel Journalism and the Negotiation of Cultural Difference

Ben Cocking

Introduction

The intention of this chapter is to examine how British travel journalism provides its readership with cultural frames of reference for different tourist settings around the world. Specifically, the focus here will be on a set of touristic experiences that have long been established in collective British imaginings of places 'out there': safari holidays in Africa. Drawing on a selection of articles from the travel supplements of British weekend newspapers, *The Sunday Times* and *The Telegraph*, I address two principal concerns. First, I interrogate representational strategies deployed in travel journalism on these two regions and how they are constitutive of broader British imaginings of Africa. Specifically, I ask in what ways the representational strategies of authenticity function in locating aspects of the past in the present. Second, it is important that these representations are considered in terms of the commercial context in which they are produced. In addressing these concerns and examining the representational strategies in play in specific articles, the intention is also to explore how this coverage is indicative of the broader cultural frames of reference through which travel journalism views its 'others'.

In Britain, the origins of this form of journalism, the travel sections of national newspapers now compete with general and specialist travel magazines and the rapidly growing travel-related Internet sites, forums and blogs. Allied with this, travel journalism necessarily involves representing 'other' peoples and 'other' places to the 'home', providing its readers with potential templates through which to experience specific locations (Spurr, 1993, p. 19). As Steward has noted, British travel journalism has historically presented its

176

readers with the world as a set of potential experiences to be chosen and consumed, by constantly asking 'Where will you go next?', they presented their readers with a set of choices through which they could express their individual tastes and preferences. (Steward, 2005, p. 52)

Travel features arguably bear an influence on the dynamic interchange between the local and the reader/tourist: it provides us with a template of where to go, what to do and how to act while we are there. In this sense, the modes of representation deployed in travel journalism are very significant, constructing the cultural frames through which the readership views and interacts with 'others'.

In representing 'others', travel journalism deploys what Fürsich and Kavoori call 'constructs of authenticity' (2001, p. 157). In this respect, a drive for the authentic – one very much akin to MacCannell's notion of the tourist desiring to 'see life as it is really lived' (1973, p. 594) – is very often a constitutive element of the travel article's narrative. In part this seems to be characteristic of the broader commercial environment of lifestyle advertising and individualized consumption, but it also often takes the form of re-imagining – of even re-enacting – older, colonialist discourses. In this way, the ebb and flow between past and present in travel journalism functions as a marker of authenticity, which, in turn, underscores the notion that its readership encounters a personalized and unique experience.

Studying travel journalism

As has been noted elsewhere in this edited collection, travel journalism has not received the attention of other areas of the profession such as political reporting (Allan, 2004, pp. 46–58). However, since the early 2000s – and largely as a result of the efforts of the editors of this collection – it is becoming an emergent area of academic interest (Hanusch, 2009, p. 625). Particularly significant in this regard is the article 'Mapping a critical framework for the study of travel journalism' (Fürsich and Kavoori, 2001). Their article points out that tourism research has tended to focus on the economic conditions of its subject, rather than 'its social and cultural role in contemporary society' (2001, p. 152). Further, in identifying tourism and the pursuit of leisure time as a significant social practice, it is suggested that travel journalism functions as a 'site where meaning is created and where a collective version of the "Other/We" is negotiated, contested and constantly redefined' (2001, p. 167). Fürsich and Kavoori outline a key dimension of research in this field, and it is perhaps not surprising to find that this emphasis on the representational characteristics of travel journalism has been the most fertile and productive form of study in this area over the past decade (see for example, Cocking, 2009; Fürsich, 2002a, 2002b; Fürsich and Robins, 2004; Hanefors and Mossberg, 2002; Mahmood, 2005; Santos, 2004a, 2004b, 2006).

As recognition of the importance of this form of journalism has become more widespread, recent work in this area has become more diversified. For example, McGaurr draws on Beck's concept of cosmopolitanism in order to examine the genre's potential for mediating environmental conflict (2009, p. 51). By contrast, Pan and Ryan (2009) make use of content analysis in order to assess the relationship between travelogues on different regions of New Zealand and perceptions based on sensory allusions. While a considerable amount of work has been done on the role perceptions, values and ethics of journalists working in areas such as political reporting, the first study of its kind to focus on travel journalists was published only in 2012 (Hanusch, 2012).

However, a focus on the representational strategies and discursive modalities of the genre has tended to predominate. This interest may reflect the literary associations of travel journalism and often calls upon theoretical concepts developed in literary studies on travel writing. In examining the representational strategies deployed in a selection of British travel journalism, the intention here is to make use of these types of analysis: Mary Louise Pratt's work *Imperial Eyes: Travel Writing and Transculturation* (1992), David Spurr's book *The Rhetoric of Empire: Colonial Discourse in Journalism, Travel Writing and Imperial Administration* (1993) and Corinne Fowler's *Chasing Tales: Travel Writing, Journalism and the History of British ideas about Afghanistan* (2007) share in a similar endeavor, identifying what Spurr refers to as 'a range of tropes, conceptual categories, and logical operations' across a range of texts (1993, p. 3). This approach facilitates the tracing of historical and imaginative trajectories. In Fowler's case this takes the form of exploring 'those notional Afghanistans that have prevailed in the popular British imagination from the early nineteenth century to the present' (2007, p. 3). In this respect, all three authors are involved in exploring the ways in which modes of representation located in specific texts might be seen as significations of broader, colonial and imperial discourses.

Similarly, one of three contextual frameworks Fürsich and Kavoori (2001) outline as applicable to the study of travel journalism is concerned with its representational characteristics. These are periodization, power and identity, and experience and phenomenology (2001, pp. 154–155). Whilst periodization is concerned with locating the cultural and social practices of travel and tourism 'within the historical development of western societies' (2001, p. 155), experience and phenomenology seeks to formulate 'typologies of tourists/tourism and tourist experiences' (2001, p. 164). The issue of power and identity is addressed through the concept of cultural imperialism and in this regard draws on Pratt's discursive approach in order to facilitate the study of 'the representational practices of travel journalism' (2001, p. 161).

Such approaches can be seen more broadly as building on Edward Said's seminal work *Orientalism* (1978/1991). Certainly, Said places a similar emphasis on identifying common signifying practices and locating them within broader colonial and imperial discursive formations (1991, pp. 2–3, 22–23, 94). This is not to suggest that Said's model provides a wholesale template for the study of representational strategies in travel journalism. It is important to acknowledge the variations and differences in colonial and imperial practices from region to region. The representational characteristics of the travel journalism studied here may share similarities with Orientalism, but they cannot be neatly housed within this discourse. Aside from issues of geographical locality, on a theoretical level, it is also important to be mindful that Said conceived of Orientalism as a largely unchanging discourse. This fixity is not only at odds with his contention that he deploys 'Michel Foucault's notion of discourse' (1991, p. 3). As critics have noted, 'unlike Foucault, who posits not a continuous discourse over time but epistemological breaks between different periods, Said asserts the unified character of Western discourse over some two millennia' (Porter, 1993, p. 152). There is a further problem: the emphasis here on locating travel journalism's discursive production in the broader context of leisure consumption and commercialization undoubtedly lies outside of the scope of Orientalist discourse. It is important, therefore, to take account of the epistemological assumptions that govern discursive formations, as well as the broader context in which they are produced. In so doing, it is possible to examine the two principal concerns of this chapter. First, what representational strategies are deployed in travel journalism and are they constitutive of a British imagining of Africa? And to what extent do these strategies make use of older, colonialist and imperialist frames of reference? Second, it is important that these representational concerns are considered in terms of the commercial context in which they are produced. Specifically, does locating the past in the present serve as a means of 'authenticating' the journalist's travel experience? How does evoking past discourses serve the modern, commercial context in which they are produced?

Analytical strategies

As exemplars for this chapter, I use a selection of recent newspaper travel articles that cover trips to Africa. The broadsheet newspaper articles under consideration are: 'Walk on the very wild side' by Kirstie Bird (*The Sunday Times*, 9 November 2008), 'The Bush Telegraph: The wonder of elephants' by Richard Madden (*The Telegraph*, 22 February 2013), 'It's time to join the trickle back to the bushveld' by Graham Boynton (*The Telegraph*, 14 July 2012), and 'Historic sites, wild beauty and an idyllic beach retreat' by Sandra Howard (*The Telegraph*, 26 January 2013). These broadsheets are at the more conservative end of the political spectrum and their readerships are from similar demographics: the aspirant, affluent middle classes who typically invest above average cultural and financial capital in activities such as travel. While this small selection of articles in no way provides the basis for an extensive empirical study, they are representative of the kind of travel journalism typically produced in British broadsheet newspapers. They do not possess attributes that mark them out as unique in any way. Indeed, the findings of this analysis would remain relatively stable and consistent were they to be supplanted by any number of other travel articles on Africa one can find in British broadsheet newspapers. In this sense, the premise of this small sample is that they are indicative of the typical frames of reference British travel journalism tends to call upon when addressing tourism in Africa. The emphasis in this chapter is on exploring discursive elements that 'involves the analysis of multiple levels of signs, symbols, and other signifiers evoked in the text' (Fürsich and Robins, 2004, p. 141). The use of discourse analysis here assumes that individual media representations tend to be based on 'entrenched and predefined ways of portraying Others' (Fürsich, 2010, p. 9). Thus, the representational characteristics of this small sample of newspaper articles speak of the broader discursive framework in which they were produced.

Representations of Africa in travel journalism

The headlines in this selection mostly play on commonly held assumptions about Africa, tapping into collective British imagination and memory of the region by drawing on imperial and colonial rhetoric. The language draws on what Jarosz, amongst others, has referred to as the metaphor of 'Africa as the Dark Continent' (1992, p. 105). This casting of Africa is the result of its historically negative articulation in

Europe and North America in broad cultural and [the circulation of] racial terms, as 'dark,' 'savage,' 'barbarous,' 'heathen,' 'uncivilized,' and 'underdeveloped,' – expressions that have had profound ramifications on the idea of Africa in Western thought. (Grinker, Lubkemann and Steiner, 2010, p. 4)

Such signifiers are particularly apparent in headlines such as 'Bush Telegraph: The wonder of elephants' (which also makes a neat play on the fact that it was published in *The Telegraph*), 'It's time to join the trickle back to the bushveld' or 'Walk on the very wild side'. Perhaps in a different area of journalism headlines of this sort might stand out as cliché laden. However, they are typical of the headlines that fill the travel sections of British broadsheets. Yet, this is not to suggest they are indicative of a particularly hackneyed or lazy form of journalism. Rather, they are evidence of a representational trope at the very core of this form of writing. In this respect, it is important not to overlook the purpose of the genre: it must – via a narrative formed out of the dynamic interchange between the commercial pressures of both the newspaper and tourism industries – convince its readership to keep reading and keep traveling. The places offered up as potential destinations must appear appealingly different, yet not threateningly so. Thus, the fact that headlines such as these are in line with the British collective imagination is, in one sense, unremarkable and typical of the genre. However, what they conjure up in our collective imagination is another matter entirely: it is one that reveals a great deal about the relationship between tourist practices and the broader discursive context. In this sense, these headlines draw on very long-standing Western cultural assumptions about Africa, assumptions that construct it as exotic, as 'other' and as pre-modern. For example, writing about 19th-century imaginings of Africa, Koivunen (2009, p. 1) notes it 'was generally referred to in Europe as an 'unknown', 'exotic', or 'dark' in everyday language'. These representational tropes occupy discursive space in today's travel journalism, but they are also characteristics inherent to Britain's colonial and imperial relationship with Africa in the past. Writing about the imaginings of Africa in the late 19th and early 20th centuries, Coombes (1994, p. 2) notes, 'the Africa that existed in the popular imagination was an ideological space, at once savage, threatening, exotic and productive ...'. This is not to suggest that a clearly defined trajectory exists, linking colonial discourse of the past to tourist experiences of the present but rather that there are echoes and traces of these earlier ways of representing Africa that act as touch stones for the British collective imagination.

Beyond the headlines of these articles it is clear that they all, to a greater or lesser extent, make use of forms of authenticity that construct Africa as pre-modern and exotic. This seems to manifest itself in the use of what might be termed 'naturalist' modes of representation, often countered by the reassuring presence of references to Western modernity. For example, Howard writes:

> The wildlife is free to roam and, as the supremely civilized rooms were dotted about the landscaped grounds, a guard was on hand to escort us to and from the main building after dark. (2013, p. 15)

Here again, our cultural understanding of Africa is so engrained that images of wild animals, safaris, and the bush immediately come to the fore. It is *precisely* what we might imagine a holiday in Africa to involve. We are assured by its presence, tantalized by the exotic, other worldliness of 'Facing up to an elephant ... my heart nearly leapt out of my mouth when we stumbled upon a herd on our first game walk' (Bird, 2008, p. 8). Yet, while accounts of such experiences function as markers of authenticity, confirming our cultural understanding of what a tourist experience such as this should involve, they are also juxtaposed to reassurances that the comforts and security of Western living present. The guard on hand to escort you to your room after dark, for example, or statements such as 'the roads are fine if somewhat potholed, the telephones work, Wi-Fi and all the other aspects of modern connectivity operate efficiently' (Boynton, 2012, p. 16). Among this selection of articles, this dual presence of the primitive unchanging past and civilized, modern present is perhaps most explicitly articulated in Madden's article on Botswana:

> And what a camp! The luxury guests enjoy here is not so much one of five-star baubles – although all creature comforts are provided for from al fresco copper baths to gourmet dining and a swimming pool – as that of visiting an eccentric (and very wealthy) friend with a country house in the bush. Built around a centuries-old termite mound in the shade of ancient jackalberry trees, guests can read in the open-fronted library as bushbucks drink from the lagoon just a few feet away. At night the campfire is lit, gin and tonics are served, hippos snort and the reed frog glockenspiel band begins its nightly jamming session. (2013, n.p.)

This juxtaposing of representations of nature with the modern conveniences of 'civilized' society functions as a means of authenticating the experience for the reader/tourist as well as reassuring them. It is also a modus operandi at the very heart of British (and other European) colonial practices: there is a long history of British travelers heading up parties of natives carrying huge quantities of luggage and bringing with them the trappings of home (Koivunen, 2009, p. 50).

The people of Africa are not absent from this vision that focuses on its wild and beautiful nature. But they are prescribed specific roles. Descriptions of people seem to place them in different facilitatory roles, and these appear to be defined largely on racial grounds. Typically, the camp guides who ensure tourists see, for example, a pride of lions, are black men. These men enable access to nature and, more than this, descriptions of their cultural heritage are very often included and serve as a means of illustrating their credentials, expertise and authenticity in providing this access to nature. This is apparent in Madden's account of the elephant handlers:

> [A]ll of whom seem as gentle and wise as the elephants themselves, tell you anecdotes from their lives with the Abu herd. Legendary Big Joe has been with them from the start and starred alongside Clint Eastwood and the original Abu himself in the 1990s film White Hunter, Black Heart. (2013, p. n/a)

The black handlers here are placed near to nature. Literally, the biological characteristics attributed to the elephants – 'gentle and wise' are infused with those who work with them. In other instances, black men are cast not only as enabling access to nature but as the means by which the tourist connects with Africa's rich cultural heritage. For example, Howard's holiday in South Africa includes a visit to Isandlwana, where on 22 January 1879 more than 1300 British soldiers died in a battle with Zulu warriors, and Rorke's Drift where 100 British soldiers defeated 4000 Zulus. Howard's guide, Mphuwa, had a personal connection to the battlefield. His

> ... great-grandfather and grandfather had fought at Isandlwana and he described taking Lord Chelmsford's great-grandson to see the memorial. 'He stood looking at it', Mphuwa said, 'then he held me in his arms and we both had a little cry'. (Howard, 2013, p. 15)

Here it is the guide's cultural heritage that brings, for the tourist, the past into the present. However, in referring to the meeting of direct descendants from both sides of the battle, it is a version of the past over which reconciliation has been adjudged to have taken place. It is, though, a form of authenticity that has been largely constructed in a reassuringly value-neutral way, unsullied by Western postcolonial guilt. Consequently, the reader is provided with an opportunity to make a personal and direct connection with the past:

> To go to Rorke's Drift the next day, seeing the site at sunset and listening to tales of supreme courage and endurance, was strange yet quite as deeply affecting [sic]. The buildings are still there, some reconstructed, but just as they were. I felt transported back. (Howard, 2013, p. 15)

Black women are less present in this selection of articles, but, again, where they appear there are echoes of older, colonial practices and representations. It is a practice very similar to the process of recalling, renewing and transforming of colonialist and imperialist tropes Pratt finds at work in contemporary travel writing (1992, pp. 221–224). There is a tendency in this selection of articles to present black women in more domestic settings. In some instances, this amounts to somewhat patronizing or 'child-like' portrayals of black women: 'Wendy, our cheerful tour-guide' or '... Grace, who served us was a tonic. At breakfast when I asked how she'd slept, she said "Like a log but I didn't end up on the fireplace"' (Howard, 2013, p. 15). Other representations of black women also place them close to nature and their cultural heritage but, in contrast to black men, do so in a domestic setting. For example, Howard (2013, p. 15) remarks on the number of traditionally styled round Zulu houses, their female guide replies: '"The Zulus believe it helps ancestors to recognize their old homes" ... "It's also for the snakes; they like to curl up in corners but with circular walls they keep on going and out the door again!"'.

The over-riding impression is that black people are encountered as workers; the guides, the drivers, the guards, the cleaners of the safari camps in which middle class, Western, predominantly white, guests stay. Yet, despite these relatively low-grade positions, there is a reverence in these narratives for their local knowledge and expertise and their cultural heritage. Indeed, as MacEachern notes in the context of black guides taking tourists on tours of South African townships: 'Quite clearly their job is one which demands that they identify, frame and interpret this crucial element of their culture and history for an audience' (MacEachern, 2002, p. 100). Black tourism workers become signifiers of unhindered access to nature and history.

In a sense, the ways in which black people are cast mirrors earlier colonial discourses (Comaroff and Comaroff, 2010, p. 32), but in this context it also functions as a means of bringing authenticity to the tourist experience. These representations help facilitate a sense of authenticity akin to MacCannell's belief in a 'touristic desire to share in the real life of the places visited' (1976, p. 96).

If black people are constructed as the facilitators of authentic tourist experiences – guides who function as 'cultural brokers' (Reisinger and Steiner, 2006, p. 482) of local knowledge – then there is a tendency here to portray white Africans as the gatekeepers of this episteme. Typically, they are mentioned as the owners of the hotels, safari camps and resorts. More than this they are presented as being the driving force behind the holiday experiences on offer. For example, Bird makes it clear within the first few lines that the concept of the 'Wildguiding adventure' in Botswana can be attributed to Clinton 'Cliffy' Philips, who runs Tuli Safari Lodge (2008, p. 8). Similarly, Howard's experience of visiting historic Zulu battlefields was the idea of the original owner of Fugitive's Drift lodge, David Rattray (2013, p. 15), who 'believed passionately that the stories of Isandlwana and Rorke's Drift should be told and preserved for posterity'. As with earlier colonial and imperial practices, white people are cast in roles that place them at the top of the hierarchy. Indeed, in Boynton's (2012, p. 17) article on Zimbabwe, there is a good deal of empathy towards white landowners who have lost land and property to 'howling mobs of so-called "war veterans"'. This empathy borders on nostalgia for pre-Mugabe models of land ownership. Here, the predominantly white lodge owners are portrayed as the guardians of the bush with conservation tourism being presented as the means of sustaining the natural environment:

> One of these farmers is Cedric Wilde, whose 25,000-acre game farm just outside Bulawayo was once a model ranch, richly stocked with rare and beautiful beasts such as sable antelope, gemsbok (imported from Namibia) and leopard. Since the farm's 'liberation', many of these have fled and been poached and the place is now in ruins. (Boynton, 2012, p. 17)

Tourist experiences

Whilst the representations of the nature and people of Africa function tend to recall the past, often evoking old, colonialist and imperialist discourses in the highly commercialized environment of today's travel journalism, they also contribute to the sense that these tourist experiences are unique and individual. Referents of the past are transferred into the present by constructing the experience as authentic and personalized.

MacCannell's concept of 'staged authenticity' (1973, 1976) is apposite here. MacCannell makes use of Erving Goffman's (1959/1990) notion of front stage to back stage as means of demarcating the social spaces of the tourist sight. MacCannell posits authenticity as a continuum, ranging from the 'false' (1973, 1976), constructed front stage, to the 'realm of "truth", "reality" and "intimacy"' of the back stage (Lau, 2010, p. 480). Whilst MacCannell conceives of the concept of 'staged authenticity' in respect of the arrangement of tourist sights, a similar continuum occurs on the textual level in travel journalism. Certainly, amongst the selection of articles examined here, the search for and description of 'back stage' experiences is a distinctive measure of a successful, and, ultimately, 'authentic' trip. To achieve this authenticity, the individualization of the experience is a central plot that plays out in two ways.

Firstly, holidaying on safari in Africa fits with a broader set of moral, political and cultural values, which draw on a mix of colonialist nostalgia and contemporary environmentalism – and the reader is asked to identify with this lifestyle choice. Secondly, within this, the reader is being presented with the possibility of a personalized holiday experience. In this sense, the naturalist representational tropes found in this selection of travel journalism not only echo and evoke the past; they are also portrayed as uniquely personalized. For example, Boynton's experience of safari lodges in Zimbabwe conveys this sense of individualized experience:

> We sat beside it [a swimming pool], about 10ft from a herd of elephants, all drinking from the pool as if it were the local water hole. Which indeed it was. I've spent a lot of time in the African bushveld, and have to say that sitting among the wild elephants as we did that morning was up there with the best wildlife experiences. (Boynton, 2012, p. 17)

Instances such as these reveal to the reader the opportunity to experience something exceptional, something momentous even. In MacCannell's terms it is the possibility, the lure of 'back stage'. The uniqueness of these experiences is premised around another commonly held cultural assumption about Africa: that its exotic distance (and perhaps expense) makes it a once-in-a-lifetime opportunity. The tourists' experiences will be cherished as rare moments in which they encounter the 'real life' of others in an unfettered return to nature. For example, Madden says of his experience of encountering elephants: 'being close to these amazing creatures has been a unique privilege' (2013, p. 2). In other instances, the sense that these experiences are not only unique but life-changing is conveyed much more dramatically. This is particularly apparent in Bird's description of her final night in Botswana. Here, there is a sense of a very British love of Africa, something which, again, is also very redolent of earlier, colonial times:

> We sat with sundowners talking about our experiences and I caught one of the guides looking at me. 'I was just watching Africa enter your soul,' he said. Normally, I wouldn't go with that kind of film-star dialogue but in Botswana, it made perfect sense. (Bird, 2008, p. 8)

Conveying a sense of authenticity and individualized experience is not limited to constructing Africa as the past in the present. In most of the articles under consideration, references to Western modernity also seem to add a further layer of individualized experience. It is as though there is an assumption that the image of Africa as pre-modern lurks in the background of the British cultural imagination and the references to Western modernity are set off against this. In this sense, they function as a means of individualizing the tourist experience. They seem to suggest that for us tourists, personally, a small slice of Western style luxury has been carved out of a pre-modern, uncivilized mass. Typically, however, luxury is not presented as contemporary and cutting-edge but established through a sense of long-standing grandeur. Here, as elsewhere, there are instances where such modes of representation draw upon the colonial past, deploying it as a marker of taste, classiness and sumptuousness while providing a reassuringly familiar context in which the reader can place the experience:

> Our room at the lodge, with its veranda and views over the historic estate, was comfortable and colonial in style; the thermos of chilled water, the barrel of meltingly light shortbread and the flowered cosy over the teapot awakened old memories. The beautiful garden and large swimming pool were an unexpected bonus and a marvelous new library overlooked the dramatic Buffalo River Gorge, the original boundary between Zululand and Natal. (Howard, 2013, p. 15)

In this way long-established signifiers of the (colonial) past are subtly balanced with markers of Western modernity. Markers of the latter appear though to be highly selected; they are not those associated with the 'hum drum' or 'daily grind' of contemporary urban life. By contrast, they are referents of 'comfort' – home cooking, books, calming and well-appointed interiors; they speak to the personalized moments we carve out of our daily lives. Put in the context of holidaying in Africa these referents are laid on just for tourists, they nourish and ensure that the exotic, other-worldliness of Africa is especially synthesized into thrilling and enchanting moments rather than triggering our deep rooted fears by manifesting itself as threatening.

Conclusions

It is clear that the commercial environment in which travel journalism is produced means that it has to be attuned to the cultural expectations and aspirations of its readers. The content of travel journalism can, at times, be cliché ridden and ethically and politically suspect. Yet, to dismiss it as such is to miss the point. Fundamentally, the significance of this form of journalism is its role as cross-cultural mediator (Hanusch, 2012, p. 683). Principally, this functions in two ways.

First, travel journalism taps into our collective cultural imagination. It draws on and, largely, reaffirms what we already know about different regions of the world. This is where the representational strategies it deploys are of interest: not so that we can dismiss, for example, the ways in which it evokes older, colonial discourses as lazy, politically suspect journalism. Rather, because finding such strategies at work is revealing of the cultural power of this form of journalism. It has long been accepted that the use of specific representational strategies in the 'hard' news of political reporting comes to bear on our political views and our understanding of politics (Allan, 2004, p. 2). It is important to accept that travel journalism similarly shapes our cultural imagination of the world. As noted above, the focus here on Africa is used in that it is indicative of the ways in which travel journalism shapes our sense of place. Focusing on travel journalism about other regions of the world would not reveal the same modes of representation, but in all likelihood would further substantiate the point that travel journalism draws on and perpetuates popular and entrenched ways of constructing other peoples and places.

Second, it is important to recognize that travel journalism has the potential to impact cross-cultural encounters and tourist practices more generally. For example, travel articles are often accompanied by tips on where to stay, where to eat and what to visit. Conceivably, then, this has the potential to function as a template on which the readers/tourists base their visit. Similar claims have been made about the influence of guidebooks, travel writing or, indeed, travel-related content in other media forms (Cocking, 2009, p. 66). What is of significance is the potential here for travel journalism to become the lens through which the reader/tourist sees parts of the world (Urry, 2002, p. 1). There is a process of perpetuation at work here; the modes of representation deployed in travel journalism affirm (to some extent at least) the cultural preconceptions of its readership. In this respect, changes on a representational level can inevitably only happen gradually. Thus, while the sample of travel journalism under consideration in this chapter was small, it is indicative of the broader representational characteristics in operation in the genre. In this sense, travel journalism can be seen as producing a degree of representational consistency. Current tourist practices of holidaying in Africa may be quite far removed from the colonial and imperial power structures of the past. However, the fact that they are evoked and reproduced in contemporary travel journalism suggests that elements of these older discourses are still present in popular imaginings of Africa.

It follows, therefore, that, if this is how Africa is understood, representational tropes that evoke colonial and imperialist signifying practices function in the context of travel journalism as a construct of authenticity. Such representational strategies – which recall characteristics of earlier 'naturalist' modes – provide the reader with the lure of authenticity, a chance to see the 'real Africa'.

While MacCannell's original conception of authenticity refers to the social practices of the tourist setting, it is particularly relevant here in that it helps provide a framework to examine representational strategies of travel journalism. It is hoped that aspects of this study might be of use to others in this field. The constructs of authenticity offer readers images that are exotic and 'other', tantalizing them with the allure of experiences far beyond the structures of everyday life at home. At the same time, such representations are countered by the reassuring presence of Western modernity and the trappings of luxury. The reader is presented with the opportunity to access an ideal pre-modern nature of Africa without having to engage in issues such as poverty, political turmoil and corruption that Africa also conjures in the popular imagination. Another dimension to this authenticity is that it draws on and reaffirms representations of Africa circulating in popular media culture: films like 'Out of Africa' or wildlife documentaries such as those made by David Attenborough.

Perhaps consideration of these representations of authenticity can be pursued, by way of a concluding observation, a little further. It seems that travel journalism presents its readers with the opportunity to undertake specific tourist experiences and these are made appealing by markers of authenticity. We are exhorted to buy into and go on particular holidays, and travel journalism provides us with a template for doing precisely that. More than this, part of the appeal of reading this form of journalism must also surely lie not in taking up one of these tourist experiences but in merely reading about them. Arguably, there is a vicarious pleasure in imagining oneself trekking with elephants in Botswana, for example. The constructs of authenticity in travel journalism also appear to operate on this level. They point the way to the actualities of specific tourist practices but they also provide their readers with a textually based experience of authenticity. The appeal of travel journalism, therefore, is that it promises authenticity on the page as much as it assures us it is 'out there'.

References

Allan, Stuart (2004) *News Culture*, 2nd edn., Maidenhead, Berkshire: Open University Press.

Bird, Kirstie (2008) 'Walk on the very wild side', *The Sunday Times*, 9 November, p. 8.

Boynton, Graham (2012) 'Its time to join the trickle back to the bushveld', *The Telegraph*, 14 July, pp. 16–17.

Cocking, Ben (2009) 'Travel Journalism: Europe imagining the Middle East', *Journalism Studies*, 10.1, pp. 54–68.

Comaroff, Jean and Comaroff, John (2010) 'Africa Observed: Discourses of the Imperial Imagination' in Grinker, Roy Richard, Lubkemann, Stephen C., and Steiner, Christopher B. (eds), *Perspectives on Africa: a reader in cultural history, and representation*, 2nd edn. Oxford: Wiley-Blackwell, pp. 31–43.

Coombes, Annie, E. (1994) *Reinventing Africa: Museums, Material Culture and Popular Imagination*, New Haven and London: Yale University Press.

Fowler, Corinne (2007) *Chasing Tales: travel writing, journalism and the history of British ideas about Afghanistan*, Amsterdam and New York: Rodopi.

Fürsich, Elfriede (2002a) 'How can global journalists represent the 'Other'?: A critical assessment of the cultural studies concept for media practice,' *Journalism*, 3.1, pp. 57–81.

Fürsich, Elfriede (2002b) 'Packaging Culture: the potential and limitations of travel programs on global television,' *Communication Quarterly*, 50.2, pp. 204–26.

Fürsich, Elfriede (2010) 'Media and the representation of Others,' *International Social Science Journal*, 61.199, pp. 113–30.

Fürsich, Elfriede and Kavoori, Anandam P. (2001) 'Mapping a critical framework for the study of travel journalism,' *International Journal of Cultural Studies*, 4.2, pp. 149–71.

Fürsich, Elfriede and Robins, Melinda B. (2004) 'Visiting Africa: constructions of nation and identity on travel websites,' *Journal of Asian and African Studies*, 39.1–2, pp. 133–152.

Goffman, Erving (1959/1990) *The Presentation of Self in Everyday Life*, London: Penguin.

Grinker, Roy Richard, Lubkemann, Stephen C., and Steiner, Christopher B. (eds) (2010) *Perspectives on Africa: A Reader in Cultural History, and Representation*, 2nd edn., Oxford: Wiley-Blackwell.

Hanefors, Monica and Mossberg, Lena (2002) 'TV Travel Shows – a pre-taste of the destination,' *Journal of Vacation Marketing*, 8.3, pp. 235–246.

Hanusch, Folker (2009) 'Taking travel journalism seriously: Suggestions for scientific inquiry into a neglected genre', *ANZCA09 Communications, Creativity and Global Citizenship*, Conference Proceedings. Available at: http://www.dev.internet-thinking.com.au/anzca_x/images/stories/past_conferences/ANZCA09/hanusch_anzca09.pdf

Hanusch, Folker (2012) 'A profile of Australian travel journalists' professional views and ethical standards', *Journalism*, 13.5, pp. 668–86.

Howard, Sandra (2013) 'Historic sites, wild beauty and an idyllic beach retreat', *The Telegraph*, 26 January, pp. 14–15.

Jarosz, Lucy (1992) 'Constructing the Dark Continent: Metaphor as Geographic Representation of Africa,' *Geografiska Annaler. Series B, Human Geography*, 74.2, pp. 105–115.

Koivunen, Leila (2009) *Visualizing Africa in Nineteenth-Century British Travel Accounts*, London: Routledge.

Lau, Raymond W.K. (2010) 'Revisiting Authenticity a Social Realist Approach,' *Annals of Tourism Research*, 37.2, pp. 478–98.

MacCannell, Dean (1973) 'Staged Authenticity: Arrangements of Social Space in Tourist Setting,' *American Journal of Sociology*, 79.3, pp. 589–603.

MacCannell, Dean (1976/1999) *The Tourist a New Theory of the Leisure Class*, California: California University Press.

MacEachern, Charmaine (2002) *Narratives of Nation Media, Memory and Representation in the Making of the New South Africa*, New York: Nova Science Publishers.

Madden, Richard (2013) 'The Bush Telegraph: The wonder of elephants', *The Telegraph*, 22 February. Available at: http://www.telegraph.co.uk/travel/activityandadventure/9879541/The-Bush-Telegraph-The-wonder-of-elephants.html

Mahmood, Reaz (2005) 'Travelling Away from Culture: the dominance of consumerism on the Travel Channel', *Global Media Journal*, 4.6, http://lass.calumet.purdue.edu/cca/gmj/sp05/graduatesp05/gmj_sp05_graduateTOC.htm, accessed 6 May 2012.

McGaurr, Lyn (2009) 'Travel Journalism and Environmental Conflict', *Journalism Studies*, 11.1, pp. 50–67.

Pan, Steve and Ryan, Chris (2009) 'Tourism Sense-Making: The Role of the Senses and Travel Journalism', *Journal of Travel & Tourism Marketing*, 26.7, pp. 625–639.

Porter, Dennis (1993) 'Orientalism and its problems', in Patrick Williams and Laura Chrisman (eds), *Colonial Discourse and Post-Colonial Theory: A Reader*. Harlow, England: Longman, pp. 150–61.

Pratt, Mary Louise (1992) *Imperial Eyes: Travel Writing and Transculturation*, London: Routledge.

Reisinger, Yvette and Steiner, Carol (2006) 'Reconceptualising interpretation: The role of tour guides in authentic tourism', *Current Issues in Tourism*, 9.6, pp. 481–98.

Said, Edward (1978/1991), *Orientalism*, London: Penguin.

Santos, Carla Almeida (2004a) 'Perception and interpretation of leisure travel articles', *Leisure Sciences*, 26, pp. 385–405.

Santos, Carla Almeida (2004b) 'Framing Portugal: representational dynamics', *Annals of Tourism Research'*, 31.1, pp. 122–38.

Santos, Carla Almeida (2006) 'Cultural politics in contemporary travel writing', *Annals of Tourism Research'*, 33.3, pp. 624–44.

Spurr, David (1993) *The Rhetoric of Empire: Colonial Discourse in Journalism, Travel Writing, and Imperial Administration*, Durham and London: Duke University Press.

Steward, Jill (2005) '"How and where to go": The role of travel journalism in Britain and the evolution of foreign tourist, 1840–1914', in John K. Walton (ed.) *Histories of Tourism: Representation, Identity and Conflict*. Clevedon, UK: Channel View Publications, pp. 39–54.

Urry, John (2002) *The Tourist Gaze*, 2nd edn., London: Sage.

11
Authorizing Others: Portrayals of Middle Eastern Destinations in Travel Media

Christine N. Buzinde, Eunice Eunjung Yoo and C. Bjørn Peterson

Introduction

Media analysts have long discussed the powerful role of media in discursively constructing realities pertaining to foreign conflicts (Price and Tewksbury, 1997). According to Tumber and Webster (2006), media are central to the debate on how to comprehend international conflict. Media representations are thought to be particularly influential when an audience has limited or no contact with the portrayed population (Fujioka, 1999). Scholars have examined the dominant media portrayal of the Middle East, which they claim is most often articulated via frames of terrorism, fanaticism, Islamist radicalism and an overall threat to Western society (Hashem, 1997; Slade, 1981). The 9/11 tragedy and subsequent US-led wars in the Middle East are currently central to the framing of this region (Steiner, 2007; Timothy and Daher, 2009). As Wang, Ding, Scott and Fan (2010) argue, 'The watershed events of September 11 terrorist attacks and the subsequent worldwide war on terrorism have exacerbated an already distorted attitude towards Muslims and the Islamic regions' (p. 118). Post 9/11, Kumar (2010), for instance, outlines five central frames used in dominant American media constructions of the Middle East: hideboundness/inflexibility, gender discrimination – male dominance/oppressed women, irrationality, violence and terrorism. The ideology underpinning these frames regards the 'West' as a leader of democracy and enlightenment and the Muslim world as mired in backwardness and intolerance (Kumar, 2010). Framing the Middle Eastern World in this way expresses the binary opposition – 'us versus them' – outlined in Said's (1979) notion of Orientalism; in his analysis of an American-produced television documentary, Said (1979) notes the absence of Muslim sources to challenge this opposition.

Tourism scholarship has also engaged in examining the unique circumstances in this politically volatile area. There exist a number of studies that focus on the Middle East from a destination image and marketing perspective and from a management standpoint (see Al-Hamarneh and Steiner, 2004; Al Mahadin and Burns, 2007; Schneider and Sönmez, 1999; Sönmez and Sirakaya, 2002; Zamani-Farahani and Henderson, 2010). Some have focused particularly on the seminal edited text by Scott and Jafari (2010), *Tourism in the Muslim World*. The prolegomenon offers details regarding the nature of the tourism industry within states like Saudi Arabia, Jordan and Iran and discusses the adverse influence of negative news coverage on the destination image of these states. As Steiner (2010, p. 185) explains, some 'Muslim destinations [within the Middle East] extensively depend on the generating markets of the Western ... world whose tourists seem to be particularly sensitive to violent political unrest' and tend to view the region homogenously and as uniformly plagued with political conflict.

Most media or tourism analyses that interrogate media coverage of the Middle East typically focus on news journalism (Santos, 2004). However, studies of traditional media content, although useful, are not sufficient, particularly considering that audience interest in 'hard' foreign news is diminishing. Consequently, many audiences receive information on foreign countries in popular media and other journalistic forms such as travel journalism (Fürsich, 2002a; Yoo and Buzinde, 2011). The US media industry produces increasing numbers of televised travelogues that offer audiences multi-sensory experiences of foreign people and places, some of which feature the Middle East. One simply has to turn to programs aired on US cable networks for proof of the popularity of televised travelogues. Much like news media, televised travelogues actively participate in the symbolic construction and dissemination of information on Middle Eastern nations. TV travel programs often combine journalistic styles of reporting – to enhance the reliability of the featured information – with entertainment strategies that engage viewers. Given the globalization of media content, audiences could include both tourists and members of the featured host nation and, in this way, televised travelogue depictions could influence both how the former perceive the latter and how the latter perceive themselves.

Since travel journalism must be 'alive to the cultural expectations and experience of its consumers', we can expect that the discourse will serve to reassure its audience of both the 'otherness' and safety of the destination (Cocking, 2009, p. 65). Given the prolonged US–Middle

East political conflict and the problematic nature of dominant US news media content on the Middle East, the proliferation of American televised travelogues, particularly those featuring the Middle East, raises certain questions: Which discursive strategies are employed in the content of televised travelogues featuring the Middle East? Moreover, as the Middle East is cast as the definitive 'Other' to western tourists (Cocking, 2009), can these shows counter the dominant US news discourse on the Middle East? Using textual analysis, this study examines the visual and verbal content produced by the American primetime televised travelogue, *No Reservations*, to understand the discursive strategies used to portray Middle Eastern states and how these strategies interact, if at all, with dominant US-based news media discourses on the Middle East.

From a sociological perspective, it is crucial to examine these discursive strategies and media representations of foreign people and places, because these shows fashion myths and expectations that affect the ways certain groups of people – locals or tourists – are perceived and interpreted (Selwyn, 1996). According to Andsager and Drzewiecka (2002) 'what tourists see, experience and learn about other societies, whose countries they visit, often depends on existing structures of pictorial representation and interpretation of cultural Others' (p. 401). Consequently, critical inquiry into these discursive practices can offer valuable insights into the 'world-shaping/world making projective authority of tourism' (Hollinshead, 2003, p. 267) that aims to 're-narrativize and remould the image of places' (Hollinshead, Ateljevic and Ali, 2009, p. 429). The role of tourism and its (dis)engagement with international conflict is vital to our understanding of the complex geopolitics that define our global village.

Tourism portrayals of people and places: the case of televised travelogues

Few studies deal with the interplay between news media and the content showcased in televised travel programs (Santos, 2005). Several studies focus on the general content of travel programs (see Dunn, 2005; Fürsich, 2002a; Jaworski, Ylänne-McEwen, Thurlow and Lawson, 2003); but research on the discursive strategies adopted by televised travelogues featuring the Middle East, and the interplay with mass media meta-narratives on the region, has remained scarce. There are some notable studies that do not focus on Middle Eastern nations but nonetheless illustrate some of the discursive practices utilized in televised travelogues. For instance, Dunn's (2005) work indicates that the content (re)produces tourism spaces through a gaze grounded in dominant ideologies that result in objectified portrayals of the Other. Similarly, drawing on the BBC's *Holiday* and the Independent Television Authority's (ITA or ITV) Wish You Were Here, Jaworski, Ylänne-McEwen, Thurlow and Lawson (2003) examine host–guest interactions. Their findings indicate that locals are given minimal roles in the narrative; their inclusion is 'limited to some expert talk, service encounters and brief phatic exchanges' (p. 158). Likewise, examining *Rough Guide*, *Lonely Planet*, and *Travelers*, Fürsich (2002a) states that, other than the typical depiction as service workers, the featured accounts do not actively involve locals in the showcasing of their nation. While she examines a drama rather than a travel program, Waade's (2011) examination of the role of landscape and setting in constructing an outsider's view of a country is also relevant. Waade (2011, p. 58) concludes that 'settings and landscapes of TV series are used to ensure realism, recognition and familiarity, and at the same time to create estrangement (with a touristic view of ... landscape and culture) and aesthetification (emphasizing the picturesque)'.

Although adopting different foci, a key commonality of this research is the argument that the depiction of the Other is entrenched in dominant ideologies and often neglects local issues as articulated, lived and dealt with *by locals* (Osagie and Buzinde, 2011; Yoo and Buzinde, 2011). The absence of the local in the constructed narratives enacts 'symbolic violence' (Spivak, 1988) and results in superficial accounts (see Fürsich, 2002a), which essentially omit 'the many resonances and initiatives [that] locality brings to the everyday complexity of our world' (Osagie and Buzinde, 2011, p. 212). Through innovative 'modes of mediation' and advanced technologies, televised travelogues offer viewers a sense of co-presence, albeit virtual, and render 'greater visibility to differences ... raising in heightened forms the problem of knowable community' (Dubey, 2003, p. 3). The ubiquitous nature of televised travelogues therefore warrants further investigation in order to understand the (re) construction of knowledge, albeit from a tourism perspective, on the region. Our goal is not to distinguish right from wrong or accurate from inaccurate, but, rather, to understand the essence of the adopted discursive strategies. Our research on this issue has the potential of augmenting tourism scholarship on a region with a thriving tourism industry despite continued political unrest (Steiner, 2010; Timothy and Daher, 2009; UNWTO, 2010).

Method

No Reservations is a US-produced, primetime, cable-televised travelogue that premiered in 2005. The program airs on *The Travel Channel*, which is owned by Scripps Networks Interactive (SNI) that is traded as SNI on the New York Stock Exchange. Scripps Networks Interactive also owns and operates HGTV, Food Network and other cable television channels (Travel Channel, 2013). For US residents, this channel is included in basic cable packages, so residents with limited viewing access (no cable) may not have exposure to the program. We focus on this particular travel program as it is part of an international network of channels and therefore reaches a wide range of audiences. In 2005, former owner and distributer Discovery Communications aired the Travel Channel around the world on the *Discovery Travel & Living Channel*. Currently SNI distributes the channel internationally, featuring numerous tourism destinations (Travel Channel, 2013). Due to its quality and popularity, *No Reservations* has been nominated for and awarded several reputed accolades including an Emmy Award for Outstanding Informational Programming and a Primetime Creative Arts Emmy Award for cinematography in the non-fiction category (Travel Channel, 2010). In addition to a focus on tourism, the program's distinct angle on gastronomy is complemented by Anthony Bourdain, the host and celebrity chef whose straightforward and opinionated voice differentiates the program from similar shows.

Middle Eastern states are showcased in a number of TV travel series; however, at the time of this investigation, which took place in 2010, *No Reservations* was the only American program that had featured more than one Middle Eastern state. The program's repertoire included episodes on Saudi Arabia and Lebanon, both of which were incorporated in this study. One of the program's trademarks is its inclusion of local co-presenters who help Bourdain interpret local culture and define the international cultural landscape.

The use of textual analysis in this study allows us to examine both how texts produce meanings and what meanings – whether latent or manifest – are embedded in the text (Hughes, 2007). A text, here, is regarded as a reservoir of cultural practices and discursive meanings based on particular cultural contexts (Real, 1996). Textual analysis is relevant to the investigation of media content as it focuses on underlying cultural assumptions, ideologies and the meanings of a text (McGuigan, 1997). This approach considers narrative media content as a site of

198 Buzinde et al.

ideological negotiation and representation of reality. Moreover, textual analysis has already been successfully utilized for critical analyses of other travel shows (e.g., Fürsich, 2009).

Analysis

The discursive strategies adopted by the televised travelogues and dominant news media discourses on the Middle East interact. Despite its location within the mainstream media institution, *No Reservations* adopted counter-frames in its depiction of Saudi Arabia and Lebanon. Each state was depicted using a distinct theme: a *9/11 theme* framed the depictions of Jeddah, Saudi Arabia, and an *Israel–Lebanon War theme* framed the depictions of Beirut, Lebanon. In both, locals were involved in the discursive co-construction of their state/city as a tourist destination – Danya, a female Saudi-American film producer (along with her family and friends) co-presents the Saudi Arabian episode, while Joe Codek, a Lebanese playwright and director (along with other locals) co-presents the episode on Lebanon.

Portrayal of Jeddah, Saudi Arabia through 9/11

The Saudi Arabian episode begins with various individuals competing for the opportunity to co-present a show featuring their home nation/city, from which Bourdain selects Danya. Danya is a resident of both the US and Saudi Arabia and is the first female filmmaker to be granted permission to film in Saudi Arabia. Intrigued by Danya's presentation, Bourdain hesitantly agrees to film an episode in Saudi Arabia:

> I think I picked Saudi because I figured that of all the choices, Saudi would be the hard thing, the most challenging thing. There were a lot of preconceptions to overcome. After all, this is where fifteen of the nineteen hijackers came from, maybe it's my sheer contrariness but I ended up going just about the last place I wanted to go. Danya challenged me to see how ordinary Saudis live their lives, feed themselves, and entertain guests.

At the outset of the episode, then, the audience is informed of Bourdain's unease due to the 9/11 tragedy. The audience is oriented to the interplay between American mainstream media discourses and the narrative produced in the televised travelogue with the mention of the hijackers. Subsequently, the episode is underpinned by tension, suspicion and both comfort and discomfort as Bourdain overcomes his preconceptions – dominant US media frames form the lens through which he deciphers the people and kingdom of Saudi Arabia. Mentioning the hijackers would likely resonate with audiences who were exposed to the 9/11 news media discourses on terrorism, and so a certain psychological proximity between Bourdain and the audience is developed; concurrently relevance and salience is articulated.

In the episode, Bourdain is guided by Danya to an historic, two-story building with a roof-top view of Mecca, where they are to prepare and enjoy a festive meal. However, all efforts towards meal preparation are halted as the hour of prayer dons. The audience is shown a thread of images depicting local Muslim devotees 'closing up shop' in response to the call for prayer. Juxtaposing dominant US media accounts, which portray the region as filled with religious fundamentalists and the reality surrounding him, Bourdain states:

> For a Westerner, it's a reminder of where you are, removed from what you see on TV, what you might already think or assume, if you can put all of that out of your head in a vacuum, it's lovely and impressive.

Despite his preconceptions, Bourdain is both pleasantly surprised and receptive to what he sees. His use of the words *lovely and impressive* certainly counters contemporary dominant media characterizations of Muslim people. Yet, despite his appreciation for this devotion, there still exists a schism brought on by his internalization of dominant tropes; he finds it difficult to accept as reality what appears before his eyes.

This is one of the first instances in the episode where Bourdain engages with the existence of other realities and worldviews and the interplay between these and his own worldview. His commentary counters dominant US media that often frame the Arab world in terms of religious fanaticism and/or humorlessness (Kumar, 2010; Merskin, 2004; Shaheen, 1984). In contrast, Bourdain offers an alternate lens through which to view the Saudis:

> I'm beginning to notice by the way, that it's not just Danya with the sense of humor around here, people actually laugh at themselves. Irony is not an unknown entity. It is a cheerful, whimsical, good humored and sophisticated atmosphere, very much at odds with the kind of humorless fanaticism I was led to expect.

He further disqualifies the image of the 'humorless' Arab while interacting with other locals, repeatedly noting their comic commentaries and laid-back nature, or *joie de vivre*.

Although the show begins with and predominantly concentrates on these two issues, it nonetheless incorporates the usual tropes of hedonism and exoticism that characterize the showcasing/promotion of tourist destinations – audiences are informed of leisure activities such as the recreational use of All-Terrain-Vehicles in the sand dunes, shopping at the mall, air-hockey, picnics and family dinners, which bridge the cultural gap insofar as they are popular pastimes of people from many nations. This strategy capitalizes on the 'familiar' and endears the place and its people to Western audiences – Bourdain even compares Jeddah to an American city, stating that, minus the alcohol and gambling, 'it looks a bit like Vegas, a desert city that can't quite grow fast enough' – and 'seek[s] to apply the values of home to give meanings to the destination abroad.' Consequently, the message is rendered relevant and salient for the Western viewing public (Dunn, 2005, p.163). Contrasts between the two states are also highlighted and, like dominant news media discourses, aim to separate the self and Other. However, unlike dominant news media discourses, the representational strategy of novelty, deployed in the episode in question, aims to lure audiences to the location by accentuating the exotic; predominantly, through unique gastronomic concoctions like lizard soup and camel stew.

Portrayals of Beirut, Lebanon through the Lebanon–Israel war

This episode commences with a discussion on the 2006 war between Lebanon and Israel. This political event is of particular relevance because the *No Reservations* crew was physically in Lebanon in 2006, attempting to film the episode, when the war between Israel and Lebanon commenced, and they were subsequently unable to complete their filming. In his opening remarks, Bourdain orients audiences to Beirut, Lebanon, as a tourist destination through a discussion of this war. Given this, his presentation incorporates issues typically not found in narratives of tourist landscapes – death, bombs, suffering, fear for one's life, anxiety, uncertainty, political unrest, shame and guilt. Some of the pictorial representations of this discussion include pre-war imagery of Beirut coupled with visuals of the war; images of bombs being dropped from fighter jets; bombed buildings; and, masses of people huddled in small shelters.

> Last time I left Beirut in 2006. It was not a happy time for me or for Beirut. I will remember looking out at the airport we flew in on and being shell bombed and rocketed and coming to the realization that wow we are not getting out of here certainly not the way we came

in ... Last time I left Beirut it was a heart break. Something I feel ashamed of, saddened by, angry about, incomplete. After two days of shooting the show in a city that felt more like Miami Beach than the center of all things wrong with the world, me and my crew found ourselves in the middle of a war. We spent eight days watching bombs and rockets fall, held up in the luxurious confines of Royale Hotel, looking down and around as the city, only recently rebuilt after years of civil war and just freed from its Syrian occupiers, was smashed back years.

The camera transitions to a gathering between Bourdain and the original crew, who reflect on their experience particularly the anxiety and uncertainty of whether they would live to tell their story. The crew expresses sentiments of anger and frustration that surfaced when they were flown out of Lebanon alongside other expatriates while locals were left behind to suffer:

Bourdain: So how do you feel to be back here?
Crewmember: It feels a little surreal. As though we're back in these haunted locations. I'm looking forward to doing a, a *No Reservations* show, a real *No Res* ... you know?
Bourdain: I want to get it right. I want to make the show we tried to make back in 2006.

Having reflected on this first journey, Bourdain orients audiences to the post-war state of the nation in 2010. Two key local figures help him recount the past and narrate the present – Ramsey, a local journalist, and Codek, a Lebanese playwright and director, who were both involved in the uncompleted filming of 2006:

Bourdain: Joe was with us right here when it happened ... Joe is remarkably more downbeat, gloomy since I last saw him. I remember you were really pessimistic right away [during the political tensions] and you said something like they will bomb us ten years back and it happened.
Joe: There was a lot of tension and they [the two nations] passed the limit and I knew there was gonna be a war, you know.
Bourdain: And Hezbollah is more powerful than ever, right?
Joe: Definitely, and I think that what made them the biggest favor is the Israeli strike.

The conversation moves beyond past accounts of war to the present situation in which Hezbollah, an extremist Shiite Muslim group, is claimed to have gained control of Lebanon. The narration of this political strife from the Lebanese perspective is key in countering the dominant (pro-Israel) perspective adopted by American news media (Sönmez, 2001; Timothy and Daher, 2009), and accomplishes a counter-framing of Lebanon that disrupts dominant ideology. This unconventional introduction to a tourist destination continues to unfold as Bourdain and Codek revisit local landmarks filmed in the pre-war episode and show sensationally dynamic images of post-war reality and how Lebanon was affected by the war. Bourdain focuses especially on the post-war involvement of Hezbollah in the reconstruction of Beirut:

> What happened after the war? Shiite neighborhoods like this one, only ten minutes from our hotel were hit hard. What happened next? Hezbollah, flush with Iranian money, they rebuilt your home, neighbor security they have it, they care for the kids, and schools they provided. They have been very, very smart enough to turn Lebanese misfortune to their own advantages. They are now more powerful, and influential than ever, with the government, and on the street.

This description illustrates the complexity and disorderliness that characterizes a society where the boundaries in binaries like oppressor/oppressed, victim/perpetrator, bad/good are blurred.

Some may contest the accuracy of Bourdain's account or critique his singular perspective, but it remains notable that the scenario created by Bourdain orients audiences to contentious issues that define Lebanese society. Mainstream media discourses on the political tension between Israel and Lebanon or Israel and the Arab world at large are partly based on accounts of the latter as a monolithic world of fanatic Muslims (Naber, 2009). Rather than perpetuate this homogenous perspective, Bourdain presents Lebanon in terms of complexity, entwinement and hybridity. For instance, he states:

> Beirut, back in the 60s, was known as the Paris of the Mediterranean. A lot of different groups, languages, interests, religions, sects, organizations, political factions, a lot of problems. But somehow it seemed for a while to be working. It's a big beautiful city with a lot going on. Two worlds: Sleek, flashy, trendy, and consumeristic: body worshiping, well dressed, chic, glittery. And ten minutes away: poor, still bomb damaged, Hezbollah, everywhere, refugee camps. Christians, Jews, Shiite Muslims, Sunni, Druids ... Gulf state money flowing in, Syrian agents, tourists, models, nightclub promoters and

DJs, Western entrepreneurs, you hear Arabic, English, French, inter-changeably, still, for whatever reason, even with all the problems, and all the terrible things that have happened here over the years, I step off the plane in Beirut and I feel strangely, inexplicably com-fortable, happy, at home.

This excerpt dismantles dominant portrayals that offer homogenous depictions of all Arabs. Showcasing the local particularities evident in Lebanon, Bourdain provides an alternate view that problematizes dominant portrayals of a static and well-defined place/people. It is also notable that the above excerpt speaks to the juxtaposition of genres involved in representing Lebanon; in addition to the typical travelogue genre, the episode incorporates genres such as television news and action/adventure, which both magnify the importance of the emergent theme. By discussing critical social issues, the episode neither romanti-cizes nor exoticizes Lebanon but, rather, aims to alert audiences to the problems that plague society from which tourism is not immune. In the excerpt below, Bourdain offers a list of activities that align people in Lebanon with many pastime behaviors of Americans; however, he continues to alert the audience to the dominant political ideology that the US imposes on Lebanon (see below his passing remark on Donald Rumsfeld's foreign policy visions):

Beirut, Lebanon, this is a country with seemingly everything. A one-time reputation as a go-to destination for the international jet-setter. Great beaches, fertile valleys, Roman ruins, a hub of the Arab world, where billboards of half clad women do not pose the problem they might elsewhere. Incredible food, good wine, amazing seafood, lamb, produce, a long tradition of nightlife, and yet they just can't seem to catch a break. It seems, on its face, a Rumsfeldian dream of the Middle East, what the deep thinkers and global strategists and smart guys say they want the Middle East to look like.

Bourdain invites the audience to view this once flourishing tourist destination, now recovering in the aftermath of war, as a space that cannot be excised from its political past and present; appreciating and understanding it requires cognizance of its politico-historical existence.

Much like the narrative created for Saudi Arabia, the Lebanon episode also commences with, and principally concentrates on, socio-political issues. This discussion is followed by the accounts of hedonism and exoticism applied to local gastronomy. Food emerges as a unique pull factor, but it is also captured in such a way as to highlight the confluence of various ethnic cuisines resulting in unique infusions that have come to be associated with modern day Lebanon. Market stalls, unique restaurant experiences and family dinners make up the gamut of gastronomic accounts featured in this episode.

Discussion

These two *No Reservations* episodes transcend the normative standards often adopted in tourism media in which local socio-political issues are typically neglected (Fürsich, 2002a). Engaging dominant American news media frames, the content of the two televised travelogues offers an alternate interpretive lens through which to understand the featured Middle Eastern countries. Socio-political issues are central to the narratives produced by the travelogues. The episodes adopt counter-narratives, which orient audiences to Saudi Arabia and Lebanon principally through two lenses – *9/11* and the *Israel–Lebanon War*. Critical film analysts explain that the measures that determine importance of an issue within televised programs are *placement* and *length* (Jamieson and Campbell, 1997). According to the authors, the typical length of a discussion on television programs ranges from a minute to two-and-a-half minutes; longer discussions are considered more important. Similarly, issues discussed towards the beginning of a program are regarded as more significant than those discussed towards the end. In the analyzed episodes, the emergent themes took up a substantive portion of the aired show: more than the *first half* of each episode and woven into the latter half. The placement and length of the emergent themes in the analyzed travelogues indicate the importance of these issues – to the presenter/producer and/or the target audience.

There are three principal strategies that aid in the discursive construction of counter-narratives within the analyzed episodes, namely: the attitude and reported experience of the presenter, the discursive construction of the Middle East as a leisurescape, and the foregrounding of the Other. Bourdain's opinionated character and authoritative voice, coupled with his unapologetically liberal perspective, form the lens through which audiences are oriented to alternative views. His pointed, comical and often politically incorrect remarks traverse boundaries as they relate to, for instance, gender, religion, money or sex. Bourdain's unrestrained character and remarks not only express the show's name, *No Reservations*, but also alert unsuspecting viewers to the fact that the show adopts a provocative and atypical approach. The articulation of the Middle East as a leisurescape also aids the discursive construction of the counter-narrative as it presents a sharp contrast to the dominant framing of Arab states as zones of war and/or terror. Audiences are informed about how to navigate the projected landscapes as foreign tourists; familiarity and novelty are built into the narrative to encourage vicarious (and perhaps physical) travel. The show does not ignore the dominant frames but instead uses them as reference points from which to orient audiences to the portrayed leisurescape.

The projected counter-narrative is further grounded by the strategic engagement of locals in the narration of their homelands. This is important principally because dominant media narratives tend to depersonalize and dissociate the Other. By contrast, the foregrounding of the Other allows for a more 'personal' connection with *a named Other* who speaks a familiar language. The show presents what Hale (2006) calls the 'authorized Other' rather than the 'insurrectionary Other.' According to Hale (2006), the 'authorized other' exhibits ideals similar to those held by Westerners, while the latter, which is central in dominant news media, is recognized but displaced and/or misplaced. The foregrounding of the 'authorized Other' establishes locals as experts on an equal level with the presenter and overcomes problematic portrayals of locals (unnamed and voiceless) in other travel programs.

From a critical perspective, these episodic themes cannot necessarily be regarded as indicators of inclusive representational practices, particularly because, as articulated by Fürsich (2003), travel programs, as entities intricately linked to commercial production, are likely to produce non-controversial and/or celebratory accounts in order to appeal to a wider audience. In this case, *No Reservations* lures audiences through its discussion of dominant narratives, however its construction of an alternate view is not merely a counter-narrative but a discursive strategy that neutralizes and/or sanitizes the controversies within the dominant content. Bourdain 'authorizes' an 'Other' through his own framing of the conflict, effectively neutralizing the potential anxiety a viewer may experience. Bourdain's 'authorized Other' is made the key protagonist in the emergent pro-diversity narrative (Hale, 2006) that facilitates the product differentiation of the featured tourism product. Ambiguity and messiness are thereby removed, allowing viewers once again to focus on the central product of the show – tourism.

Inclusivity, as it relates to these shows, can be further elucidated through the concept of multiculturalism, which can be thought of as an approach to acceptance of difference or as a critical interrogation of the West in relation to the rest (Buzinde, Santos and Smith, 2006).

According to critical educational pedagogues, the former dimension denotes a celebration of cultural diversity and is referred to as liberal multiculturalism (Rezai-Rahsti, 1995). By contrast, the latter dimension is referred to as an anti-racist pedagogy and aims to disrupt the status quo and to investigate the historical and ideological (re)production of cultural differences that exacerbate social inequalities (Ng, 1995; Rezai-Rahsti, 1995). Liberal multiculturalism emphasizes the just and/or equal treatment for all, while anti-racist multiculturalism focuses on how difference is articulated and aims to deconstruct systems of oppression. These approaches converge in their claims that identity and difference are non-dynamic concepts, facilitating political support for oppressed groups and for those mobilizing efforts for equal recognition and respect.

Liberal multiculturalism is more relevant when explaining the strategies adopted in these episodes. Specifically, efforts were taken to account for equality in the content by including the Other and the construction of a discourse of sameness founded on the notion of common humanity (West, 1994). However, in order to mobilize the argument of sameness, the element of identity-as-difference is evoked and the narrative cements separations between ethnic/racial identities. In other words, the elaboration of identity is stated in terms of what people do not have in common, as opposed to what they do hold in common. The affirmation of sameness is offered only in relation to the more fundamental difference. In addition to the accentuation of difference, the liberal multicultural approach also facilitates the silencing or omission of certain issues. Left out are issues accentuating differences that may not be overcome by a discourse of sameness.

Consequently, the travelogue's liberal multicultural approach fails to offer a comprehensive account regarding the historic-political relationship between the Middle East and the West, a history that continues to define contemporary relationships between the two geopolitical locales. Nevertheless, it is important to note that the program's engagement with socio-political issues that are otherwise often excised from tourism discourse is a vital step towards making tourism more relevant to societal issues beyond commercial/managerial foci that have traditionally defined the industry and academic field.

Conclusion

The interweaving of dominant American media discourses in the narration of Middle Eastern nations as tourism destinations illustrates the intertextual nature of media portrayals. The deployed discursive strategies went beyond the surface gaze to offer nuanced depictions of socio-political issues that define tourism locations. Audiences were directed towards that which is contentious, uncomfortable, problematic, complex and unresolved; this strategy presented a politicized gaze that encouraged audiences to acknowledge rather than escape the everyday problematic situation. Furthermore, the shows attempted to reduce what Spivak (1988) refers to as 'symbolic violence' by grounding the narrative in local perspectives with the active involvement of local co-presenters. Given this approach, one can regard the televised travelogue as a virtual, intercultural meeting space, which offers an opportunity 'to transform our way of understanding the world, by critically questioning, challenging, and problematizing... [international conflicts] and, as a result, broadening our awareness and approach' (Agathangelou and Ling, 2009, p. 86). Nevertheless, the travel shows have to be understood within a specific economic model. Their need to attract advertisers and audiences as programs on a lifestyle channel limits how Otherness and political conflict can be articulated. Local Others are presented within a framework of liberal multiculturalism while downplaying historical and ideological (re)production of cultural differences that exacerbate social inequalities (Ng, 1995; Rezai-Rahsti, 1995). From this perspective, the content furthers a liberal multiculturalist agenda that allows for the Other to be commodified for tourist purposes.

This chapter presents an initial attempt to examine the link between travel media, American mainstream news frames and global politics in the creation of meanings attributed to tourist destinations in the Middle East. It demonstrates one way in which tourism plays a key role in creating awareness of 'our mutual embeddedness' as well as the 'entwinement of [our] multiple worlds' (Agathangelou and Ling, 2009, p. 86). As tourism media continue to articulate spaces of international political conflict, it is important for scholars to further examine whether the 'messiness' that characterizes society is neglected in favor of an aestheticized portrayal. Critical analyses of media content have to continue to evaluate the potential of counter-narratives in travel journalism: The goal is to overcome sanitized portrayals that promote escapism and to adopt nuanced portrayals that offer crucial lessons to humanity.

Acknowledgment

The authors are grateful to the Department of Recreation, Park and Tourism Management (RPTM), at the Pennsylvania State University for funding this project through the Faculty Start-up Fund.

References

Agathangelou, Anna and Ling, L.H.M (2009) *Transforming world politics: From empire to multiple worlds*, New York, NY: Routledge.

Al-Hamarneh, Ala and Steiner, Christian (2004) 'Islamic tourism: Rethinking the strategies of tourism development in the Arab world after September 11, 2001', *Comparative Studies of South Asia, Africa and the Middle East*, 24.1, pp. 173–82.

Al Mahadin, Saba and Burns, Peter (2007) 'Visitors, visions and veils: The portrayal of the Arab world in tourism advertising', in: Rami Daher (Ed), *Tourism in the Middle East*, Clevedon: Channel View Publications, pp. 137–60.

Andsager, Julie and Drzewiecka, Jolanta (2002) 'Desirability of differences in destinations', *Annals of Tourism Research*, 29.2, pp. 401–21.

Buzinde, Christine N., Santos, Carla A., and Smith, Stephen L.J. (2006) 'Ethnic representations: Destination imagery', *Annals of Tourism Research*, 33.1, pp. 707–28.

Cocking, Ben (2009) 'Travel journalism: Europe imagining the Middle East', *Journalism Studies*, 10.1, 54–68.

Dubey, Madhu (2003) *Signs and cities: Black literary postmodernism*, Chicago, IL: University of Chicago Press.

Dunn, David (2005) 'Venice observed: The traveler, the tourist, the post-tourist and British television', in Adam Jaworski and Annette Pritchard (eds) *Discourse, Communication and Tourism*, Clevedon: Channel View Publication, pp. 98–120.

Fujioka, Yuki (1999) 'Television portrayals and African-American stereotypes: Examination of television effects when direct contact is lacking', *Journalism and Mass Communication Quarterly*, 76.1, pp. 52–75.

Fürsich, Elfriede (2002a) 'Packaging culture: The potential and limitations of travel programs on global television', *Communication Quarterly*, 50.2, pp. 204–26.

Fürsich, Elfriede (2003) 'Between credibility and commodification: Nonfiction entertainment as a global media genre', *International Journal of Cultural Studies*, 6.2, pp. 131–53.

Fürsich, Elfriede (2009) 'In defense of textual analysis', *Journalism Studies*, 10.2, pp. 238–52.

Hale, Charles. R. (2006) *Mas que un Indio: Racial ambivalence and neoliberal multiculturalism in Guatemala*, Santa Fe: School of American Research Press.

Hashem, Mahboub (1997) 'Coverage of Arabs in Two Leading US Newsmagazines: Time and Newsweek', in Yahya Kamalipour (Ed.) *The US Media and the Middle East*. Westport, CT: Praeger, pp. 151–62.

Hollinshead, Keith (2003) 'Symbolism in tourism: Lessons from Bali 2002-lessons from Australians dead heart', *Tourism Analysis*, 8.2, pp. 267–95.

Hollinshead, Keith, Ateljevic, Irena and Ali, Nazia (2009) 'Agency-worldmaking authority: The sovereign constitutive role of tourism', *Tourism Geographies*, 11.4, pp. 427–43.

Hughes, Peter (2007) 'Text and textual analysis', in Eoin Devereux (Ed), *Media Studies: Key Issues and Debates*, London: Sage, pp. 249–82.

Jamieson, Kathleen and Campbell, Karlyn K. (1997) *The interplay of influence: News, advertising, politics, and the mass media*, London: Wadsworth Publishing.

Jaworski, Aam, Ylänne-McEwen, Virpi, Thurlow, Crispin & Lawson, Sarah (2003) 'Social roles and negotiation of status in host-tourist interaction: A view from British television holiday programmes', *Journal of Sociolinguistics*, 7.2, pp. 135–63.

Kumar, Deepa (2010) 'Framing Islam: The resurgence of orientalism during the Bush II era', *Journal of Communication Inquiry*, 34.3, pp. 254.

McGuigan, Jim (1997) *Cultural methodologies*, London: Sage.

Merskin, Debra (2004) 'The construction of Arabs as enemies: Post-September 11 discourse of George W. Bush', *Mass Communication & Society*, 7.2, pp. 157–75.

Naber, Nadine (2009) 'Transnational families under siege: Lebanese Shi'a in Dearborn, Michigan, and the 2006 war on Lebanon', *Journal of Middle East Women's Studies*, 5.3, pp. 145–74.

Ng, Roxanna (1995) 'Teaching against the grain: Contradictions and possibilities', in Roxanna Ng, Joyce Scane and Patricia Staton (Eds), *Anti-Racism, Feminism and Critical Approaches to Education*, Toronto: OISE Press, pp. 129–52.

Osagie, Iyunolu and Buzinde, Christine N. (2011). 'Culture and postcolonial resistance: Antigua in Kincaid's A Small Place', *Annals of Tourism Research*, 38.1, 210–30

Price, Vincent and Tewksbury, David (1997) 'News values and public opinion: A theoretical account of media priming and framing', *Progress in Communication Sciences*, 45.2, pp. 173–212.

Real, Michael R. (1996). *Exploring media culture: A guide*, London: Sage.

Rezai-Rahsti, Goli (1995) 'Multicultural education, anti-racist education, and critical pedagogy: Reflections on everyday practice', in Roxanna Ng, Joyce Scane and Patricia Staton (Eds), *Anti-Racism, Feminism and Critical Approaches to Education*. Toronto: OISE Press, pp. 3–19.

Said, Edward (1979) *Orientalism*, New York: Vintage.

Santos, Carla A. (2004) 'Framing Portugal: Representational dynamics', *Annals of Tourism Research*, 31.1, pp. 122–38.

Santos, Carla A. (2005) 'Framing analysis: Examining mass mediated tourism narratives', in Ritchie, Brent W., Peter M. Burns, and Catherine A. Palmer (Eds) *Tourism Research Methods: Integrating Theory with Practice*, Cambridge: CABI Publishing, pp. 149–62.

Schneider, Ingrid and Sönmez, Sevil (1999) 'Exploring the Touristic Image of Jordan', *Tourism Management*, 20.4, pp. 539–42.

Scott, Noel and Jafari, Jafar (2010) *Tourism in the Muslim world: Bridging tourism theory and practice*, Bingley: Emerald Books.

Selwyn, Tom (1996) *The tourist image: Myths and mythmaking in tourism*, New York: Wiley.

Shaheen, Jack G. (1984) *The TV Arab*, Bowling Green, OH: Bowling Green State University Popular Press.

Slade, Shelley (1981) 'The image of the Arab in America: Analysis of a poll on American attitudes', *Middle East Journal*, 35.2, pp. 143–62.

Sönmez, Sevil (2001) 'Tourism behind the veil of Islam: women and development in the Middle East', in Giörgos Apostolopoulos, Sevil Sönmez and Dallen Timothy (Eds) *Women as Producers and Consumers of Tourism in Developing Regions*, Westport, CT: Praeger, pp. 113–42.

Sönmez, Sevil and Sirakaya, Ercan (2002) 'A distorted destination image? The case of Turkey', *Journal of Travel Research*, 41.2, pp. 185–96.

Spivak, Gayatri (1988) 'Can the subaltern speak?', in Carry Nelson & Lawrence Grossberg (Eds), *Marxism and the interpretation of culture*, Urbana: University of Illinois Press, pp. 271–313.

Steiner, Christian (2007) 'Political instability, transnational tourist companies and destination recovery in the Middle East after 9/11', *Tourism and Hospitality Planning & Development*, 4.3, pp. 167–88.

Steiner, Christian (2010) 'Impacts of September 11: A two sided neighborhood effect?', in Noel Scott and Jafar Jafari (Eds) *Tourism in the Muslim World*, Bingley: Emerald Books, pp. 181–204.

Timothy, Dallen and Daher, Rami (2009) 'Heritage tourism in Southwest Asia and North Africa: Contested pasts and veiled realities', in Dallen Timothy and Gyan P. Nyaupane (Eds) *Cultural heritage and tourism in the developing world: A regional perspective*, London: Routledge, pp. 146–64.

Travel Channel (2010) '*Anthony Bourdain No Reservations: About the show.*' Available at http://www.travelchannel.com/TV_Shows/Anthony_Bourdain/About_The_Show/Anthony_Bourdain_No_Reservations/ (accessed 3 December 2011).

Travel Channel (2013) '*About Us*'. Available at http://www.travelchannel.com/about/about-us/ (accessed 27 September 2013).

Tumber, Howard and Webster, Frank (2006) *Journalists under fire: Information war and journalistic practices*, London: Sage.

United Nations World Trade Organization (UNWTO) (2010) *Tourism barometer*, Madrid, Spain: UNWTO.

Waade, A.M. (2011) 'BBC's Wallander: Sweden seen through British eyes', *Critical Studies in Television: The International Journal of Television Studies*, 6.2, pp. 47–60.

Wang, Zhuo, Ding, Peiyi, Scott, Noel and Fan, Yezheng (2010) 'Muslim tourism in China', in Noel Scott and Jafar Jafari (Eds) *Tourism in the Muslim world: Bridging tourism theory and practice*, Bingley: Emerald Books, pp. 107–19.

West, Cornel (1994). 'The new cultural politics of difference', in Cameron McCarthy and Warren Crichlow (Eds) *Race, Identity and Representation in Education*, New York: Routledge, pp. 11–23.

Yoo, Eunice E. and Buzinde, Christine N. (2011) 'Reception theory and the interpretation of televised travel shows: The case of the Middle East', *Annals of Tourism Research*, 39.1, 221–42.

Zamani-Farahani, Hamira and Henderson, Joan C. (2010) 'Islamic tourism and managing tourism development in Islamic societies: The case of Iran and Saudi Arabia', *International Journal of Tourism Research*, 12.1, pp. 78–89.

Part IV
(Dark) Histories, Sustainability, Cosmopolitanism: The Old and New Politics of Travel Journalism

12
Representations of Interconnectedness: A Cosmopolitan Framework for Analyzing Travel Journalism

Wiebke Schoon

Introduction

The aim of this chapter is to introduce and operationalize the socio-logical concept of cosmopolitanism as an overarching framework for research on travel journalism. This framework places the analysis of travel journalism in the emerging research field of cosmopolitan media studies and in the wider context of transcultural journalism studies. Moreover, it offers fruitful ideas for further empirical analysis and contains normative elements that point to the desirable potential of the genre. The chapter begins by outlining the role of journalism as both a potentially globalizing actor and a nation builder. The sociological concept of cosmopolitanism is then introduced, in order to enhance the theoretical debate on the relation between globalizing and national-izing notions, and to reflect on functions of journalism that go beyond the formation of political public spheres. After sketching the discrete dimensions of the approach and outlining existing research on media and cosmopolitanism, I argue that travel journalism is a valuable object of study to complement conceptual and empirical work on media and cosmopolitanism. Research on media and cosmopolitanism has thus far been dominated by a pity and compassion paradigm and overtly focused on disaster and crises reporting (e.g. Chouliaraki, 2006; Cottle, 2009a). Taking travel journalism into account widens the perspective; everyday stories about the world are included and the research potential of the cosmopolitan approach is exploited more fully. To exemplify the ben-efits of looking at travel journalism from a cosmopolitan perspective, I refer to the research area of journalism and Europeanization. Building on existing research on travel journalism and on methodological ideas derived from the cosmopolitan approach, I develop a framework for the analysis of the genre and illustrate it by presenting a research design for a study on German travel journalism.

Journalism in processes of
globalization – nationalization – Europeanization

One of the contemporary challenges faced by journalism research is that of capturing journalism's ambivalent role as an agent of both globalization and nationalization. In research dealing with the interplay of media and globalization, journalistic content is generally seen as a mediator and indicator of international and transnational processes (Cottle, 2009b). Since news media are regarded as emissaries of a global public sphere (e.g. Volkmer, 2003), the emphasis is on mediated global interconnectedness and on the question of how media contribute to a sense of global belonging. In accordance with this perspective, the concept of 'global journalism' has been introduced; it emphasizes journalistic representations of 'the world as a single place' (Berglez, 2008, p. 848). Other scholars have emphasized journalism's role in the development and construction of the nation-state – both as an imagined community and as a political entity – in which news media are important mediators in processes of political deliberation (see Roosvall and Salovaara-Moring, 2010).

Amongst journalism researchers, scholars concerned especially with journalism in and about Europe have investigated the contradiction between transnationalization and national differentiations in media content. Several research projects have been triggered by the observation that, despite intensified economic and political integration in the common framework of the European Union, nationally bounded political public spheres seem to remain within the member states (e.g. AIM Research Consortium, 2007; Örnebring, 2009). In accordance with the foundation of democratic theory, journalistic media have been conceptualized as mediating between the political institutions of the EU and the citizens of Europe. This research reveals the ambivalent finding that a transnational European political public sphere is forming, but that 'national public spheres persist and nations partly retain influence' (Hepp and Wessler, 2009, p. 291). The national segmentation of the European public sphere is explained by pointing to journalistic practices of nationalization that '[refer] to the journalistic practices of embedding foreign issues in the context of one's own nation' (Lingenberg, Möller and Hepp, 2010).

While the nation and its constructing practices seem to have an ongoing and unquestioned status within journalistic discourse, the nation has also been used as an unquestioned category within theoretical reflections, empirical investigations and conceptual interpretations within large segments of media and communications studies. Even when the so called 'global-comparative turn' (Wahl-Jorgensen and Hanitzsch, 2009, p. 6) came about during the 1990s, large segments of journalism research kept a cross-national comparative outlook, differentiating between forms of journalistic content along traditional domestic-foreign lines (for further discussion of this aspect see Rantanen, 2010; Berglez, 2008). In addition, existing empirical research is overly focused on political news journalism, thereby establishing a conception of journalism that is solely concerned with the construction of political public spheres (for example, Wessler et al., 2008). Researchers tend to under-theorize or disregard transnationalizations of cultural life-worlds and social practices, and their associated representations of interconnectedness in journalistic content.

The next section will introduce the sociological concept of cosmopolitanism as an approach that enables the reflection of the outlined ambivalences by theorizing the relation between globalizing and nationalizing notions, and widening the research perspective towards journalistic representations of mundane transnational/ transcultural connections.

New cosmopolitanism as a sociological concept

Without doubt, cosmopolitanism is a contested term. Since there is 'no uniform interpretation of cosmopolitanism in the literature' (Roudometof, 2005, p. 116), the term reveals little about the analytical approach. Whereas the concept originates in the realm of philosophy and political theory, it is by now discussed and rearticulated in diverse disciplinary contexts (see Delanty, 2012). This chapter concentrates on the concept as it has been elaborated by Ulrich Beck and others (for example, Delanty, 2009), whose outspoken aim is 'to turn the often-misunderstood concept of cosmopolitanism from its philosophical head onto its social scientific feet' (Beck, 2011, p. 1346). The main point and common thread of this new cosmopolitanism lies in the fact that the dualities of the global and the local, the national and the international, have lost their sharp contours. As a consequence, the national and nation-centric analytical concepts lose their explanatory power, and new forms of conceptual and empirical analysis are required. As an analytical concept, cosmopolitanism is a continuation and specification of the term and concept of globalization. Following Beck (2006), the concept is used in three different dimensions: (1) as a theoretical perspective, (2) as a description of empirical practices and processes that cross and transcend national boundaries (cosmopolitanization) and (3) as an ethical and moral dimension. These components are analytically discrete but interrelated. To clarify this, the dimensions are sketched out in the following sections.

In the first dimension the emphasis lies on being analytically sensitive to the interpenetration of local, regional, national and global contexts, without taking one context or the other for granted. Beck (2006) develops a perspective called methodological cosmopolitanism, which arises from a critique of the so-called methodological nationalism, an outdated analytical outlook that regards 'the nation-state as a "self-evident point of departure"' (Levy cited in Beck, 2006, p. 33). From a cosmopolitan analytical angle the continuing significance of the national dimension is not denied, it is just not taken for granted; it is explored how nationality is constructed, how it is transformed and how it is interrelated with other dimensions. In this sense, national, regional and local contexts are not juxtaposed with the global; rather 'cosmopolitanization should be chiefly conceived of as globalization from *within*' (Beck and Sznaider, 2006, p. 9, emphasis in original). Thus, a cosmopolitan perspective can be employed for focusing on national or local settings and to search for 'transnational, translocal, glocal or global-national structures and patterns of relations' (Beck, 2006, p. 92). The cosmopolitan outlook is established by being sensitive to interpenetrations, to permeations and redefinitions of boundaries (Beck, 2006, p. 92). Moreover, from a cosmopolitan perspective it is not reasonable to distinguish between national and international realms, because, according to Beck and Sznaider (2006, p. 14), 'the world is generating a growing number of such mixed cases, which make less sense according to the "either/or" logic of nationality than to the "both/and" logic of the cosmopolitan vision'.

This 'growing number of mixed cases' refers to the empirical analytical dimension called '*really-existing cosmopolitanization*' (Beck, 2006, p. 19, emphasis in original). This process is linked to transnational and transcultural practices that result in a pluralization of social life-worlds, where national references become less relevant or more heterogeneous. Existing territorial units and social identifications are not only interconnected, but also change over time. In this context, the ongoing process of Europeanization is regarded as a regional case of cosmopolitanization (Beck, 2006, pp. 163–177; Beck and Grande, 2007; Delanty, 2009, pp. 200–249).

Other processes of cosmopolitanization refer to the side effects of actions, decisions or processes, which are not cosmopolitan in a positive or normative sense. The focus lies on the thematization of forced inter-dependencies of the so-called 'world risk society' (Beck, 1999, 2009). In this sense, cosmopolitanization results from reactions to the unin-tended side effect of global risks like environmental and financial crisis as well as terrorist threats (Beck, 2011). Beck puts this aspect at center stage in his work, and most of the studies inspired by his theoretical work deal with this aspect as well.

However, within this dimension, more banal and mundane forms of cosmopolitanization – linked to processes of consumption, popular cul-ture and cultural events – are taken into account. This is where media- and tourism-related practices are brought up for consideration (Beck, 2006, pp. 40–44). Beck recognizes the developments linked to mobil-ity, international tourism and transnational reporting as indicators of banal cosmopolitanization (Beck, 2006, p. 93). John Tomlinson (1999, p. 200), under the label of banal cosmopolitanism, similarly men-tioned 'the penetration of our homes by media and communication technology ... increased mobility and foreign travel'. Ulf Hannerz (2005, p. 207) states that, due to increased mobility, including 'labor migra-tion, tourism, backpacking, pilgrimages and student exchanges', more and more people are being taken 'out of their local habitats' and 'the social bases of cosmopolitanism are expanding'. In addition to experi-encing increased physical mobility, more and more people's 'horizons and imagined worlds have been affected by new media engagements und new consumption patterns' (Hannerz, 2005, p. 207).

The notion of banal cosmopolitanization also points to the fact that, as an analytical concept, cosmopolitanism can refer to people who are not aware that they are cosmopolitan, or even deny that they so: '"Cosmopolitanization" in this sense means latent cosmopolitanism, *unconscious* cosmopolitanism, *passive* cosmopolitanism' (Beck, 2006, p. 19, emphasis in original). These forms of cosmopolitanization are 'infiltrating the world of nation-states from below and transforming it from within' (Beck, 2006, p. 20).

The explicitly moral dimension of cosmopolitanism lies in 'the aware-ness of a global sphere of responsibility, the acknowledgement of the otherness of others and non-violence' (Beck, 2002, p. 36). In the cos-mopolitan ideal, cultural others are recognized as different and equal; furthermore, the concept is linked to people having specific as well as multiple identities. National or other traditional forms of belonging are not denied. Instead, the concept points to internal transformations of social spaces and to the relevance of mediation between different cultures, communities and lifestyles. Beck (2002, p. 35) defines the notion of a 'dialogical imagination' as a central characteristic of cosmopolitanism in its moral dimension and elsewhere adds that

> this involves two things: on the one hand, situating and relativizing one's own form of life within other horizons of possibility; on the other, the capacity to see oneself from the perspective of cultural others and to give this practical effect in one's own experience through the exercise of boundary-transcending imagination. (Beck, 2006, p. 89)

In this moral dimension, which is also emphasized in Gerard Delanty's 'cosmopolitan imagination' (2009), the concept most obviously differs from other concepts of globalization and transnationalization; it is more demanding with regard to the interactive and transformative dimension of cultural encounters. Cosmopolitan processes in this sense include a critical and reflective moment that points to the central role of communication and collective identity narratives.

Cosmopolitanization and the media – mediated cosmopolitanism

The notion of imagination is fundamental in this context. It refers to 'the way in which societies symbolically constitute themselves' (Delanty, 2009, p. 14), and is linked to (mass) media. Whereas Benedict Anderson (1983) has illustrated how national newspapers have played a crucial part in imagining the nation, Beck claims that nowadays all media are involved in 'processes of transnation-building' (Beck in Rantanen, 2005, p. 255). Beck (2011) exemplifies this by referring to the above-mentioned global interdependency crisis and refining 'cosmopolitanism as imagined communities of global risks' (see also Beck and Levy, 2013). Moreover, the question of whether passive and unconscious forms of cosmopolitanization become conscious and reflexive, thus leading to a (potentially ethical) cosmopolitan outlook on the world, is related to media coverage and usage. Beck recognizes that the mixing of cultures has been the rule rather than the exception throughout history, but adds, 'what is new is not the forced mixing but awareness of it ... its reflection and recognition before a global public via the mass media' (Beck, 2006, p. 21). The emerging field of cosmopolitan media studies has mainly focused on risks, and has strongly emphasized the moral component. Most attention has been paid to theoretical reflections on 'media and morality' (Silverstone, 2007) and on the question of whether media representations of crises and disasters can lead to compassion for distant others and cosmopolitan empathy (Chouliaraki, 2006; Robertson, 2008; Kyriakidou, 2009; Pantti, 2009; Ong, 2009; Cottle, 2009a, Pantti, Wahl-Jorgensen and Cottle, 2012). Most recently, Beck and Levy (2013, p. 20) have reflected on this emerging research field; they criticize relevant parts of it for reverting to a problematic binary of national discourses and cosmopolitan ideas. Instead, they argue that 'cosmopolitanization itself is a constitutive feature of the reconfiguration of nationhood' (p. 5). This point is similarly recognized by Alexa Robertson (2008, p. 23), who states that 'it is in national settings that cosmopolitan sentiments are fostered'. Her empirical work on 'mediated cosmopolitanism' (2010) provides a comprehensive perspective, by investigating television news narratives and focusing on 'the steady drip of images and stories that could shape understandings of the world in unspectacular, and thus often unnoticed, ways' (Robertson, 2010, p. 16).

Combining the cosmopolitan approach with research on travel journalism

Adding the analysis of travel journalism to the conceptual and empirical work on media and cosmopolitanism is one way to overcome the so-far dominant pity-and-compassion-framework and the focus on news journalism. By taking into consideration journalistic representations in travel journalism, we can tackle the question of how banal forms of cosmopolitanization are actually negotiated in the news. Travel journalism creates certain maps of the world that imply time-and-space contextualized outlooks on nationality, interconnections and globality. It remains to be investigated empirically whether these outlooks are parochial, or whether they are inclined to raise awareness of boundary-crossing and boundary-transcending interconnections. As previous studies have shown, research on travel journalism is a particularly interesting case, because the genre plays a growing role in the representation of various places, both near and far, and the mediatization of cultural spheres and encounters (Fürsich and Kavoori, 2001; Hanusch, 2010, 2011, 2014). While travel journalism may not be as rooted in the nation-based paradigm as political news journalism, it is clear that the genre is nevertheless steeped in the ambiguities of representing how places and people of different parts of the world are interrelated and a tendency to stereotype and essentialize cultural groups (Cocking, 2009; Fürsich, 2012; Santos, 2004, 2006). This tension within the genre can be suitably reflected from a cosmopolitan perspective, because the question of how difference is negotiated and mediated lies at the core of the concept.

On a more general level, the cosmopolitan approach provides a theoretical underpinning for integrating content beyond the news in the analysis of journalism. The cosmopolitan approach reminds us that journalistic content is not regarded exclusively as either an immanent aspect of political deliberation or a mediator in a world of crisis and disasters, but also as an agent and mediator of the diverse – sometimes even banal or mundane – processes of cosmopolitanization. By investigating the representation of banal cosmopolitanization included in (travel) features, we are thus able to enhance the social and cultural perspective in journalism studies and to take a closer look at links and distinctions between the formation of political and cultural public spheres. This corresponds with Ulf Hannerz' claim that infotainment feature stories are more likely to take world interconnectedness into account and to 'make distant places and people less one-dimensional, more complex, than they may be in much hard news reporting' (2004, p. 33). Social, cultural and entertaining aspects of journalistic stories might raise awareness and exercise imaginations that are less bound to the local life-worlds of the recipients and political territories, and eventually contribute to a deeper sense of connectivity, thus facilitating transcultural understanding. To shift the empirical and conceptual focus from outstanding events and their representation in breaking news to this kind of long-form journalism is a way of taking a closer look at journalistic representations of 'the emergent rather than the emergency' (Hannerz, 2004, p. 229).

These conceptual reflections on integrating the analysis of travel journalism in a cosmopolitan framework can be exemplified by research on journalism and Europeanization. As indicated in the beginning of this chapter, the ambivalence of transnationalizing and nationally domesticating notions is paradigmatic in this realm; however, most research in this area has focused on news journalism and the EU as a political entity. As a consequence, we know little about the distinctions made when referring to the *EU and its political institutions* on the one hand and to *Europe as a broader arena of complex socio-cultural relations* on the other (Baisnée, 2007; Golding, 2007; Kunelius and Heikkilä, 2007). In addition, there is a lack of research on the variety of imaginaries represented in entertaining forms of journalism (Örnebring, 2009). By looking at Europeanization from a cosmopolitan perspective with special emphasis on banal forms of cosmopolitanization, we can widen the scope of research to include other meanings affiliated with Europe. For example, a relevant dimension of Europeanness is linked to what is called *freedom of movement* and having the *right to travel*. EU-wide surveys and sociological studies on 'what citizens mean by feeling European' (Bruter, 2004) show that traveling, studying or working within another EU or non-EU country is a vital part of feeling European (see also Pichler, 2009). These border-crossing practices are neither uniquely European nor totally new developments within Europe, but the recent intensity of mobility is significant. Traveling as a transnational practice within and beyond Europe is linked to economical, legal and political measures taken by EU institutions, and its members or cooperating states. Examples of this are the implementation of the Euro as a currency in 18 EU states, the creation of the Schengen area with its common visa policy and, on a more symbolic level, the similarly designed EU-passports for the 28 EU member states.

Moreover, the European Commission acknowledges tourism to be 'an important means of promoting Europe's image in the world' (European Commission, 2013). The importance of tourism is even mentioned in the Lisbon Treaty (Article 195). These measures have an important impact on sociocultural practices, encounters and experiences that go beyond the narrowly defined political or economic dimension. These practices lead to transnational and transcultural relations within people's own life-world that can neither be captured with a nation-centric nor with an EU-centric perspective. The cosmopolitan perspective seems suitable to grasp the dynamics that are connected to traveling as the social practice of *doing Europe*. By looking at references to Europe and the portrayal of European subregions within travel journalism, the focus shifts from state-derived views of Europeanization to other forms of *feeling European* as it does from hard news reporting to more in-depth reporting. Empirical open questions based on these considerations are: How are tourism-related processes, places and encounters represented? Is the ongoing process of actually existing cosmopolitanization in the touristic practice also portrayed journalistically and do these portrayals offer cosmopolitan(ized) outlooks?

A cosmopolitan perspective on German travel journalism

In the following, I illustrate the contribution and potential of the cosmopolitan approach by outlining the research design of a study on German travel journalism that is based on the aforementioned theoretical considerations. This framework is part of my PhD thesis, which investigates the representation of (trans)national interconnectedness and the discourse on Europe in German travel journalism from 1979 to 2010. Objects of analysis are the travel sections of *Die Zeit* and the *Frankfurter Allgemeine Zeitung* (*FAZ*). This narrow focus on national elite print media is chosen because this media segment holds a particularly privileged position in the media discourse, and its representations can be regarded as establishing powerful patterns of knowledge. Both newspapers have been publishing weekly travel sections since the 1950s, which are among the most prestigious travel sections in Germany (Kleinsteuber and Thimm, 2008, p. 180; Vereinigung Deutscher Reisejournalisten, 2013).[1]

By investigating several sample units of travel sections that were published in the two national newspapers between 1979 and 2010, using quantitative and qualitative content analysis, I ask first what destinations are represented in the material and how this might have changed over the course of time. The longitudinal research design enables sensitivity to transformations that might have occurred in the coverage. In this first step I maintain the national analytic outlook (that is, categorizations consist of the world region, subregion, country of destination), (a) in order to keep the results comparable to existing research (Hanusch, 2011, 2014) and (b) to add a longitudinal perspective to these results. A diachronic comparison of the sample units will reveal whether the coverage is getting more diverse in terms of representing various world regions and a larger variety of countries. In other words, an increasing geographic diversity can be regarded as a basic indicator for the cosmopolitanization of the coverage and a prerequisite for a cosmopolitan outlook.

In order to go beyond nation-based differentiations, I also categorize how these regions are referred to in terms of their territorial localization. I code the two most dominant localizations within the prominent textual and visual elements of the articles and differentiate between national localization, border regions, supranational regions, subnational regions, global cities/capitals, continents and the world as a whole. Looking at the different sample units makes it possible to test whether the territorial variety (as another basic indicator of cosmopolitanization of the coverage) has increased, or whether national localizations have decreased over the course of time. It also helps to explain whether certain world regions are seen differently in terms of territorial references than others.

A major inspiration drawn from the cosmopolitan approach is to take a more nuanced look at representations of interconnectedness. This aspect is operationalized by gathering the segments of the prominent article elements[2] that deal with interconnections that do not fit in one-dimensional national contexts, and systematizing these segments according to the framework shown in Table 12.1. The first two categories grasp those elements that deal with binational or multinational interconnections, for example portraying two or more countries with central references to the national territories or citizens. That is, the outlook is a national one, but references and/or relations between two or more countries are portrayed. A typical example for the representation of multinational interconnections would be a picture showing three or more different national flags. The third category aims at capturing those textual and visual elements that refer to various (at least two) localities within different national territories, but are not framed in a national outlook. An example for this category is the visualization of a feature story about a cyclist who crosses the Alps from Munich to Venice; the caption refers to regions in southern Germany, Austria and Italy without mentioning the names of the respective countries (this story was published in *Die Zeit*, 3 October 1980). The fourth category is intended to gather those elements that represent transcultural interconnections that take place within one locality. This category can be illustrated by a story about the biggest Asian community in Europe, which is to be found in Paris (published in *Die Zeit*, 11 March 1994, and titled: 'Chinatown an der Seine'). In addition to these interconnections all explicit references to the world as a whole, the globe and globalization are accumulated.

Table 12.1 Conceptual framework to systematize journalistic content that transcends the one-dimensional national context

Category	Explanation
binational	references to two countries / 'national outlook'
multinational	references to three or more countries, 'national outlook'
transnational	two or more localities within different national territories, no or not a central national outlook
inner globalization	transnational interconnections within one locality / territorial unit
explicit notions of globality	explicit references to the world / the globe / globalization

For all segments the interconnections are also coded according to whether they are relate to tourism activities, and if not, how they are otherwise contextualized. In the more detailed analysis I look more closely at what kind of interconnections occur within the sample, whether there are differences in the sample units over time and whether there are particular patterns in the represented world regions. Generally speaking, an increase of interconnections or a decline in the nationally framed interconnections can be regarded as a further indicator of cosmopolitanization within the coverage.

Another point of interest is whether the prominent article elements contain textual or visual segments that explicitly refer to a certain nation or the inhabitants of a certain nation, or if nationality itself is an explicit theme. A closer look at the representations and narrations of nation and nationality gives insights into whether and how certain notions of nationhood are 'essentialized' (Fürsich and Kavoori, 2001, pp. 158–159), or whether and how the coverage of national notions and references has changed over time. In addition to the common focus on travel journalism as a genre that represents *foreign* destinations, the cosmopolitan analytic perspective also encourages us to take domestic stories into account and look for representations of interpenetrations, permeations and redefinitions of boundaries. In my study, domestic stories are those that deal with articles about Germany. Since reunification in 1990, German national identity has undergone major transformations, which makes domestic stories particularly interesting.[3]

Stimulating ideas that derive from the ethical and moral dimension of the cosmopolitan approach point in two directions: to the descriptive level and to the normative level. The descriptive level centers on the question of whether the stories open up 'a space of *dialogical imagination*' (Beck, 2006, p. 89, emphasis in original) and promote a boundary-crossing sphere of responsibility. In this context the textual and visual depiction of tourists and locals, their interaction and the way they are being *heard* in the stories is of special interest. Here we can see another obvious junction where cosmopolitan media studies and existing research on travel journalism meet, since the question of who is given a voice is a salient topic in both fields. Research on travel journalism that critically reflects on the genre's potential for fostering intercultural understanding is concerned with the presence of local voices (Fürsich, 2002, pp. 77–78; Hanusch, 2011, p. 28). Likewise, Silverstone (2007, p. 80) explores 'the emergence of the mediapolis as a space for multiply mediated voices' on a more abstract level; Hannerz (2004, pp. 138–143) reflects on 'voices heard and not heard', and Robertson (2010) investigates empirically which kinds of people are interviewed or quoted in television news.

In addition to asking what parts of the world are represented in the genre, and whether international news flows and the destinations referred to in travel sections follow the same lines (Hanusch 2014), the cosmopolitan perspective encourages us to investigate whether places and people are portrayed in a similar manner and, more specifically, how (relationships to) various kinds of *others* are represented. The question arises whether travel feature stories live up to their potential to build a basis for *thick cosmopolitanism* by being, for example, less elite-focused than hard news journalism, instead including more ordinary voices and everyday stories. In order to pursue this aspect it seems suitable to combine quantitative and qualitative approaches. In my study, I operationalize this by further scrutinizing those segments that have been categorized under the code interconnections (see Table 12.1) and by investigating selected subsamples more deeply. In particular, I put a special focus on the way the EU and Europe or Europeanization are referred to, in order to gain insight into how the internal transformations within Europe and the external transformations in the relation to other parts of the world are represented or constructed in the sample. Particularly with regard to the enlargement process and eastward shift of the EU, it will be interesting to find out whether the travel stories portray Europe as uniform or if there is space for the plurivocality of 'multiple Europes' (Delanty, 2009, p. 10) and the emergence of a 'cosmopolitan Europe' (Beck and Grande, 2007). Salient questions that arise in this context are: How are the destinations portrayed in terms of cultural proximity or distance? What narrative means are used to construct Europeanness? Are places and people in Eastern Europe portrayed in a less distant and more familiar manner over time? To what extent are distinctions between European and non-European spheres drawn and how this has changed?

Other topics that relate to the descriptive level of the ethical dimension of the cosmopolitan approach refer to critical contextualizations of tourism (ethical/responsible/ecotourism; see McGaurr, 2010) and to the way interdependency risks are referred to. Ensuing questions in this context relate to who is depicted as responsible for the reported problems, whether possible routes to action are presented and whether the voices that are included in the stories are conveyed as moral voices that raise critical and reflexive issues (Pantti, 2009, pp. 98–101; Robertson, 2008, p. 15).

In addition to describing the way cosmopolitanization is represented in travel sections and the questions concerning whether these representations are likely to raise awareness of boundary-crossing interconnections, or even of boundary-transcending responsibilities, the moral ideal of the cosmopolitan approach also adds a normative dimension. In this sense, the approach offers a tool for critique of current media practice. It enriches the argument that journalism bears the responsibility for mediating transcultural encounters and constructing imaginaries of near and distant cultural spheres in a respectful and sensitive manner. This approach builds an ethical foundation for the way transcultural relations and diversity should ideally be represented from a theoretical and scholarly point of view.

Conclusion

In this chapter I have connected research on travel journalism to the sociological concept of cosmopolitanism. The contribution of this approach for the analysis of travel journalism is to place research on the genre in a comprehensive analytical framework and to provide robust theoretical underpinnings for 'taking travel journalism seriously' (Hanusch, 2009, p. 1). On the methodological level, the cosmopolitan perspective encourages us to take a more nuanced look at the representation of transnational interconnectedness and global interdependency. It encourages us to refine the categorizations of journalistic representations of boundary-crossing interconnectedness and simultaneously to remain open to localizing and nationalizing references.

Moreover, the cosmopolitan approach can be regarded as an overarching framework that supports the exploration not only of the complex empirical reality but – with regard to its moral/ethical dimension – also of the desirable potential of travel journalism. On a more general level, the combination of travel journalism studies and the cosmopolitan sociological perspective enriches current debates concerning the role of journalism as both a potentially globalizing actor and a nation builder. In this sense the concepts aims to transgress the traditional domestic/ foreign dichotomy that has hitherto led to the common but anachronistic differentiation in national and international coverage. From the perspective of cosmopolitan media studies, the main benefit of taking travel journalism into consideration is to overcome the focus on distant suffering, crisis and disasters.

Notes

1. I focus on editorial parts of the sections and mostly on feature stories. The systematically drawn sample comprises 84 sections of each newspaper and 973 articles altogether.
2. Prominent elements in this study include headline, teaser, first paragraph, other highlighted elements within the article, as well as maps, pictures and captions. These are looked at in the first steps of analysis as described here. Subsequently, subsamples of whole articles will be taken into consideration.
3. Special editions of the German travel magazine, Merian, that were published after the fall of the Berlin wall are part of the collection of the German Historical Museum in Berlin.

References

AIM Research Consortium (ed.) (2007) *Reporting and managing European News. Final report of the project 'Adequate information management in Europe' 2004–2007*, Bochum: Projektverlag.

Anderson, Benedict (1983) *Imagined communities: Reflections on the origin and spread of nationalism*, London: Verso.

Baisnée, Oliver (2007) 'The European Public Sphere Does Not Exist (At Least It's Worth Wondering...)', *European Journal of Communication*, 22.4, pp. 493–503.

Beck, Ulrich (1999) *World risk society*, Cambridge: Polity Press.

Beck, Ulrich (2002) 'The Cosmopolitan Society and Its Enemies', *Theory, Culture & Society*, 19.1–2, pp. 17–44.

Beck, Ulrich (2006) *The cosmopolitan vision*, Cambridge: Polity Press.

Beck, Ulrich (2009) *World at risk*, Cambridge: Polity Press.

Beck, Ulrich (2011) 'Cosmopolitanism as Imagined Communities of Global Risk', *American Behavioral Scientist*, 55.10, pp. 1346–61.

Beck, Ulrich and Grande, Edgar (2007) *Cosmopolitan Europe*, Cambridge: Polity Press.

Beck, Ulrich and Levy, Daniel (2013) 'Cosmopolitanized Nations: Re-imagining Collectivity in World Risk Society', *Theory, Culture & Society*, 30.2, pp. 3–31.

Beck, Ulrich and Sznaider, Natan (2006) 'Unpacking cosmopolitanism for the social sciences: A research agenda', *British Journal of Sociology*, 57.1, pp. 1–23.

Berglez, Peter (2008) 'What Is Global Journalism? Theoretical and empirical conceptualisations', *Journalism Studies*, 9.6, pp. 845–58.

Bruter, Michael (2004) 'On What Citizens Mean by feeling "European": Perceptions of News, Symbols and Borderless-ness', *Ethnic and Migration Studies*, 30.1, pp. 21–39.

Chouliaraki, Lilie (2006) *The spectatorship of suffering*, London: Sage.

Cocking, Ben (2009) 'Travel Journalism: Europe imagining the Middle East', *Journalism Studies*, 10.1, pp. 54–68.

Ong, Jonathan C. (2009) 'The cosmopolitan continuum. Locating cosmopolitanism in media and cultural studies', *Media, Culture & Society*, 31.3, pp. 449–66.

Cottle, Simon (2009a) *Global crisis reporting: Journalism in the global age*, Maidenhead and New York: Open University Press.

Cottle, Simon (2009b) 'Journalism and Globalization', in Karin Wahl-Jorgensen and Thomas Hanitzsch (eds), *The Handbook of Journalism Studies*. New York: Routledge, pp. 341–56.

Delanty, Gerard (2009) *The cosmopolitan imagination: The renewal of critical social theory*, Cambridge: Cambridge University Press.

Delanty, Gerard (ed.) (2012) *Routledge handbook of cosmopolitanism studies*, New York: Routledge.

European Commission (2013) *Tourism. EU Policy background*. Available at: http://ec.europa.eu/enterprise/sectors/tourism/background/index_en.htm.

Fürsich, Elfriede (2002) 'How Can Global Journalists Represent the "Other"?', *Journalism: Theory, Practice and Criticism*, 9.3, pp. 57–84.

Fürsich, Elfriede (2012) 'Lifestyle Journalism as Popular Journalism', *Journalism Practice*, 6.1, pp. 12–25.

Fürsich, Elfriede and Kavoori, Anandam (2001) 'Mapping a Critical Framework for the Study of Travel Journalism', *International Journal of Cultural Studies*, 4.2, pp. 149–71.

Golding, Peter (2007) 'European Journalism and the European Public Sphere. Some Thoughts on Practice and Prospects', in Hans Bohrmann and Elisabeth Klaus and Marcel Machill (eds), *Media industry, journalism culture and communication policies in Europe*. Köln: Halem, pp. 25–40.

Hannerz, Ulf (2004) *Foreign news. Exploring the world of foreign correspondents*, Chicago: University of Chicago Press.

Hannerz, Ulf (2005) 'Two Faces of Cosmopolitanism: Culture and Politics' *Statsvetenskapelig Tidskrift*, 107.3, pp. 199–213.

Hanusch, Folker (2009) 'Taking travel journalism seriously: Suggestions for scientific inquiry into a neglected genre.' *Paper presented at the annual conference of the Australian and New Zealand Communication Association (ANZCA)*, Brisbane, 8–10 July.

Hanusch, Folker (2010) 'The Dimensions of Travel Journalism: Exploring New Fields for Journalism Research beyond the News', *Journalism Studies*, 11.1, pp. 68–82.

Hanusch, Folker (2011) 'Representations of Foreign Places outside the News: An Analysis of Australian Newspaper Travel Sections', *Media International Australia*, 138, pp. 21–35.

Hanusch, Folker (2014) 'The geography of travel journalism: Mapping the flow of travel stories about foreign countries', *International Communication Gazette*, 76.1, pp. 47–66.

Hepp, Andreas and Wessler, Hartmut (2009) 'Political Discourse Cultures: Explaining the Segmented Europeanisation of Public Spheres' (English abstract of the article 'Politische Diskurskulturen–Überlegungen zur empirischen Erklärung segmentierter europäischer Öffentlichkeiten, pp. 174–97), *Medien und Kommunikationswissenschaft*, 57.2, p. 291.

Kleinsteuber, Hans J. and Thimm, Tanja (2008) *Reisejournalismus. Eine Einführung. 2., überarbeitete und erweiterte Auflage* [*Travel Journalism. An Introduction, 2nd edn.*], Wiesbaden: VS Verlag für Sozialwissenschaften.

Kunelius, Risto and Heikkilä, Heikki (2007) 'Mainstream Journalism. Problems and Potential of a European Public Sphere (EPS)', in AIM Research Consortium (eds), *Reporting and managing European News. Final report of the project 'Adequate information management in Europe' 2004–2007.* Bochum: Projektverlag, pp. 45–77.

Kyriakidou, Maria (2009) 'Imagining Ourselves Beyond the Nation? Exploring Cosmopolitanism in Relation to Media Coverage of Distant Suffering', *Studies in Ethnicity and Nationalism*, 9.3, pp. 481–96.

Lingenberg, Swantje, Möller, Johanna and Hepp, Andreas (2010) '"Doing Nation": Journalistische Praktiken der Nationalisierung Europas', Edited by 140 TranState Working Paper. Bremen. Available at: http://www.sfb597.uni-bremen.de/pages/pubApBeschreibung.php?SPRACHE=en&ID=181.

McGaurr, Lyn (2010) 'Travel Journalism and Environmental Conflict: A Cosmopolitan Perspective', *Journalism Studies*, 11.1, pp. 50–67.

Örnebring, Henrik (2009) 'Introduction: Questioning European Journalism', *Journalism Studies*, 10.1, pp. 2–17.

Pantti, Mervi (2009) 'Wave of Compassion: Nationalistic Sentiments and Cosmopolitan Sensibilities in the Finnish Press Coverage of the Tsunami Disaster', in Ullamaija Kivikuru and Lars Nord (eds), *After the Tsunami: Crisis Communication in Finland and Sweden.* Göteborg: Nordicom, pp. 83–105.

Pantti, Mervi, Wahl-Jorgensen, Karin and Cottle, Simon (2012) *Disasters and the media*, New York: Peter Lang.

Pichler, Florian (2009) 'Cosmopolitan Europe. Views and identity', *European Societies*, 11.1, pp. 3–24.

Rantanen, Terhi (2005) 'Cosmopolitanization – now!: An interview with Ulrich Beck', *Global Media and Communication*, 3.1, pp. 247–63.

Rantanen, Terhi (2010) 'Methodological Inter-Nationalism in Comparative Media Research. Flow Studies in International Communication', in Anna Roosvall and Inka Salovaara-Moring (eds), *Communicating the nation: National topographies of global media landscapes.* Göteborg: Nordicom, pp. 25–39.

Robertson, Alexa (2008) 'Cosmopolitanization and Real Time Tragedy: Television News Coverage of the Asian Tsunami', *New Global Studies*, 2.2, Article 3.

Robertson, Alexa (2010) *Mediated cosmopolitanism: The world of television news*, Cambridge, UK; Malden, MA: Polity Press.

Roosvall, Anna and Salovaara-Moring, Inka (eds) (2010) *Communicating the nation: National topographies of global media landscapes*, Göteborg: Nordicom.

Roudometof, Victor (2005) 'Transnationalism, Cosmopolitanism and Glocalization', *Current Sociology*, 53.1, pp. 113–35.

Santos, Carla Almeida (2004) 'Framing Portugal: Representational Dynamics', *Annals of Tourism Research*, 31.1, pp. 122–38.

Santos, Carla Almeida (2006) 'Cultural Politics in Contemporary Travel Writing', *Annals of Tourism Research*, 33.3, pp. 624–44.

Silverstone, Roger (2007) *Media and morality: On the rise of the mediapolis*, Cambridge: Polity Press.

Tomlinson, John (1999) *Globalization and culture*, Chicago: University of Chicago Press.

Vereinigung Deutscher Reisejournalisten (2013) *Columbus Reiseteil-Preis 2012*. Available at: http://vdrj.de/columbus-preise/columbus-reiseteil-preis/2012/uebersicht. [Association of German Travel Journalists, Columbus Travel Section Award 2012]

Volkmer, Ingrid (2003) 'The global network society and the global public sphere', *Development*, 46.1, pp. 9–16.

Wahl-Jorgensen, Karin and Hanitzsch, Thomas (2009) 'Introduction: On why and how we should do journalism studies', in Karin Wahl-Jorgensen and Thomas Hanitzsch (eds), *The Handbook of Journalism Studies*. New York: Routledge, pp. 3–16.

Wessler, Hartmut, Peters, Bernhard, Brüggemann, Michael, Kleinen-von Königslow, Katharina and Sifft, Stefanie (2008) *Transnationalization of public spheres*, Basingstoke and New York: Palgrave Macmillan.

13
Your Threat or Mine? Travel Journalists and Environmental Problems

Lyn McGaurr

Introduction

It is no secret that the frames of travel journalism tend to be anchored in the worldviews of its producers and readers (Hanusch, 2011; Santos, 2004). Yet if we expand our interest in cosmopolitanism beyond the mediation of cultural diversity, to the media's engagement with politics and the environment, we may find the genre a site with more to offer the public sphere than banal globalism (Szerszynski and Urry, 2002). Sociologist Ulrich Beck has long expressed an interest in the catalyzing contribution the media can make to cosmopolitanization. He believes that news reports of risks that transcend national borders can result in the emergence, 'more or less involuntarily', of 'a pressure to cooperate' (Beck, 2011, p. 1353) with distant others whether or not one feels 'at home in the world' (Hannerz, 2006). The study presented in this chapter asks whether there might be a similar 'community-shaping power of global risks' (2011, p. 1352) inherent in travel journalism. This question is important, because the genre's 'tacit allegiance' to the tourism industry (Fürsich and Kavoori, 2001), and its reputation for misrepresenting distant cultures (see, for example, Cocking, 2009; Daye, 2005; Santos, 2004; Voase, 2006), have so far tended to discourage scholars from considering its cosmopolitan promise.

This study draws on in-depth interviews with international travel journalists to explore factors that might influence the genre's ability to contribute to the emergence of imagined cosmopolitan communities of risks. In the following sections, I introduce the theory of cosmopolitanism and discuss its relevance to the environment, tourism and travel journalism. I then describe the qualitative methods of in-depth interviews and thick description. In the final sections, I present and discuss my results before venturing some tentative conclusions. I find that the genre can sometimes accommodate cosmopolitan concern for destinations it simultaneously praises as places worth visiting, because such concern can provide evidence of marketable editorial independence while continuing to promote high-end tourism. Environmental discourses of this kind in travel journalism may contribute to the emergence of a cosmopolitan outlook, which Beck (2006) defines by its public reflexivity. However, they do not necessarily indicate a weakening of the structural bonds between mainstream media and the global tourism industry. This imposes significant limitations on the cosmopolitan promise of the genre.

Mediated cosmopolitanism and the environment

Beck's (1992, 2006, 2009, 2011; Beck and Levy, 2013) theory of risk society and his approach to cosmopolitanism are provocative for the prominence they afford the media and environment. Beck contends that the media can shock people out of their complacency about a range of risky by-products of industrialization, such as greenhouse gases and radiation. In his view, even conflicts over whether distant populations should have a say in the fate of a country's rainforests – which he considers to be global resources – 'perform an integrating function in that they make clear that cosmopolitan solutions have to be found' (2006, p. 23). As Anderson observes, 'Beck is right to recognise that perceptions of risk are selective and different environmental issues have varying degrees of cultural potency and mediagenic dying trees and seals, for example, allow us to glimpse the bigger picture' (2000, p. 96).

Szerszynski and Urry have found that multiple mobilities, including imaginative travel via the media, 'may provide the context in which the notion of universal rights, relating not only to humans but also to animals and environments, comes to constitute a framing for collective action' (2006, p. 117). A moral environmental citizenship may be acquired, whereby 'the local becomes experienced in a different way, one in which a certain abstraction informs the very perceptions of the particular – an abstraction that makes possible the critical judgment necessary to citizenship' (Szerszynski, 2006, pp. 86–87). Holton also sees what might be described as an affective cosmopolitics already evident in modern environmentalism, 'which has arisen as a response to environmental challenges that cannot be resolved on a national basis within single countries, and which draws emotional power from images of Planet Earth under imminent threat of ecological crisis' (2009, p. 5). In Beck's

(2011) hypothesis, by contrast, cosmopolitanization does not begin with empathy or the desire to make the world a better place, but with risks revealed and the self-interested need for action recognized. Writing with co-author Daniel Levy, he contends that 'new collectivities' arising this way 'are neither traditional nor voluntary. Instead they are affiliations imagined under conditions of interdependencies imposing collective constraints' (Beck and Levy, 2013, p. 23). The authors explain:

> Ultimately the political and cultural salience of these cosmopolitan affiliations depends on how risks are mediatized and consumed as habituated practices. Hence sociability is not established under conditions of united interpretations but as a result of shared attentiveness to global risks. (Beck and Levy, 2013, p. 23)

It is a challenging, unsentimental vision that inverts the traditional argument that political cosmopolitanism will emerge from the spread of morality-driven cosmopolitanism (Beck, 2011, p. 1348). Rather, imagined communities of global risks and associated interdependencies are credited with the capacity to promote a cosmopolitan 'politicization and establishment of norms' (2011, p. 1353).

Tourism, travel journalism and the environment

As cultural mediators, tourism public relations practitioners and travel journalists help shape the consumption patterns of tourists by positioning knowledge about, and experience of, distant places as a sign of social distinction (see Urry, 2002, pp. 43, 79–81 regarding tourism choices). One of the dimensions of cultural cosmopolitanism exploited by tourism operators and travel media is what Calhoun describes as 'consumerist cosmopolitanism' – a 'soft' version 'packaged for consumer tastes' (Calhoun, 2002, pp. 888–89). He is pessimistic about the potential of 'consumerist' cosmopolitanism to promote cosmopolitan democracy:

> Food, tourism, music, literature, and clothes are all easy faces of cosmopolitanism. They are indeed broadening, literally after a fashion, but they are not hard tests for the relationship between local solidarity and international civil society. (Calhoun, 2002, p. 889)

Cosmopolitans who conform to Calhoun's description of 'consumerist' do not necessarily have a desire to immerse themselves in other cultures to the extent of acquiring cultural competence; consumerist cosmopolitans are of interest to tourism marketers as a distinctive demographic simply because they are likely to be more sophisticated, independent, objective and, therefore, demanding consumers (Cannon and Yaprak, 2002). International travel journalists help construct the tourist gaze for consumerist cosmopolitans by writing for publications owned by large media institutions (see Urry, 2002). In so doing, most will encounter the representations of other institutions – government tourism offices and the commercial tourism sector.

Yet, even as travel journalists are constructing distant destinations for their readers, they have the option of shifting their point of view to a perspective MacCannell (2011, p. 210) calls the reflexive 'second gaze'. The second gaze does not leave responsibility for its own construction to 'the corporation, the state and the apparatus of tourism representation' (2011, p. 210). Rather, aware that something is being concealed, it 'looks for the unexpected, not the extraordinary ... to open a window in structure' (2011, p. 210). One such 'gap in the cultural unconscious' enabling travel journalists to 'glimpse the symbolic in action' (MacCannell, 2011, p. 210) may be evidence that distant peoples are grappling with environmental problems not so different from their own. Travel journalists who take the opportunity to convey this to their readers step outside the usual bounds of their professional routines by representing not only the 'goods' traditionally associated with tourism consumption but also examples of what Beck (2009) describes as 'bads' – the dangerous side effects of industrialization.

Although the United Nations World Tourism Organization (UNWTO) pays tribute to the potential for tourism to have positive social and cultural effects, its focus is overwhelmingly on the industry's economic benefits to communities. To this end it promotes sustainable tourism (UNWTO, 2011) as 'an engine for development through foreign exchange earnings and the creation of direct and indirect employment' (UNWTO, n.d.). Yet, tourism itself does not come without environmental costs. Many of these costs are ostensibly local, but there is one to which the sector makes an overtly global contribution: climate change. Tourism is responsible for approximately five per cent of greenhouse gas emissions (UNWTO and United Nations Environment Programme (UNEP), 2008, p. 13), and transport accounts for around three-quarters of the sector's CO_2 emissions, with aviation the major component (UNWTO and UNEP, 2008, p. 9). As the UNWTO and United Nations Environment Programme warned in 2008, environmental costs threaten to severely constrain the developmental promise of the industry (UNWTO and UNEP, 2008, p. 27).

Non-government organizations like Ethical Traveler in the United States (established in 2002) and Tourism Concern in the United Kingdom (established in 1988) perform the affective cosmopolitics of much modern environmentalism (see Holton, 2009, p. 5). By drawing on the work of climate scientist Mike Hulme, Urry and Larsen (2011, p. 101) argue that interconnections between 'increasingly cosmopolitan' travel, media, science and NGO social practice are helping citizens understand the global nature of climate change. Thus, Urry and Larsen are prepared to posit two possible futures for tourism, should forecasts of severe oil depletion, climate change and water shortages prove accurate. In one scenario, dramatic rises in security risks and breakdowns in communication and transport networks make most tourism infeasible, but in the other, eco-responsibility becomes widespread, '*de*-globalizing the tourist gaze except through virtual travel on the internet' (2011, p. 235, original emphasis).

In Australia's poorest state – the island of Tasmania – the natural environment has been a source of conflict for more than four decades. In the late 1970s and early 1980s, for example, an enormous campaign by the environment movement to prevent the state government's Hydro-Electric Commission from damming the wild Franklin River attracted national and international attention. When the state government lost this battle as a result of the Australian government's constitutional obligations as a signatory to the UNESCO World Heritage Convention, it decided to capitalize on the publicity the environmental movement's campaign had brought the island's natural environment (Evers Consulting, 1984). Promoting Tasmania as natural also enabled government and industry to make a virtue of the state's competitive disadvantage of being a vast distance from the population centers of the northern hemisphere. Government and business could now represent that distance as having quarantined the island from many of the environmental problems experienced elsewhere on the planet. As time passed, however, NGOs began arguing that the government tourism office's promotion of Tasmania as wild, beautiful and protected was incompatible with continuing old-growth logging. In the first decade of the current century, travel features as well as news articles containing criticisms of the island's forestry practices started appearing in some of the most high-profile newspapers and magazines in Britain and the United States.

Using the Tasmanian case as an exemplar of an international tourism destination promoting itself as natural while experiencing environmental conflict, my study sought answers to three related questions about travel journalism and cosmopolitanism:

1. What factors influence travel journalists' decisions to report or ignore environmental problems?
2. Do some travel journalists see a role for themselves in building connections between their readers and distant communities?
3. What factors influence travel journalism's ability to contribute to the emergence of imagined cosmopolitan communities of risks?

Method

The following study is largely based on in-depth interviews with 11 journalists who visited Tasmania and published travel features about it in Britain or the United States between 2000 and 2010 inclusive.[1] My research also draws on interviews I conducted with some of the journalists' named sources, as well as government and non-government tourism public relations practitioners and a range of other relevant individuals. Texts I consulted as part of an associated textual and contextual analysis included the articles about Tasmania published by the interviewed journalists, other travel articles by these journalists, two other samples of international travel journalism about Tasmania, government tourism and branding strategies, parliamentary reports, local newspaper articles, tourism brochures, annual reports and consultants' reports. In gathering and analyzing my data, I took the anthropological approach of creating what Geertz (1973) describes as thick descriptions. Thick description is a particular form of ethnography that attempts accurately to represent actors' meanings by exploring the social context in which those meanings are negotiated (Anderson, 1997, p. 191). When interpretive research of this kind is used by sociologists, it is concerned to gain knowledge of the complexities of social practice, including those aspects that are not on public display. As Flyvbjerg explains, it does this by 'gradually allowing the case narrative to unfold from the diverse, complex, and sometimes conflicting stories that people, documents, and other evidence tell them' (Flyvbjerg, 2001, p. 86). My in-depth interviews with journalists and other actors were a means by which to collect and analyze first-hand experiences (Rubin and Rubin, 1995) in context, so that the more correct meaning might be drawn from a multiplicity of possible meanings.

Travel journalism and distant threats

The pressure to avoid controversy

Genre protocols favoring celebratory, consumer-oriented information often play an important part in determining whether or not environmental problems will be mentioned in international travel articles. Such decisions are influenced by professional acculturation as well as personal understandings of what constitutes tourism – a point well illustrated in the example of Jamie Doward, a staff journalist on London's *Observer* newspaper. Doward's travel article about Tasmania appeared in the *Observer* in April 2004, less than a month after an article about a proposed boycott of travel to the state in protest over its logging practices appeared in his paper's sister publication, the *Guardian* (Fickling, 2004). Even so, Doward (2004) did not mention Tasmania's forestry conflict or recall it when speaking to me in 2009 (pers. comm., 26 August). His visit to Tasmania had been as part of a highly managed group media tour hosted by the government tourism office (Doward, 2009, pers. comm., 26 August). As a staff news journalist, he only occasionally wrote travel journalism, and when commenting generally on his decisions about content he attributed them to the needs of his audience. His explanation of his decision not to refer in his text to another matter of cosmopolitan concern – a contagious facial tumor afflicting the Tasmanian devil – reveals his understanding of 'what you're supposed to be focusing on' (Doward, 2009, pers. comm., 26 August) as a journalist writing for the *Observer*'s travel section – particularly in terms of a difference between news and travel journalism:

> I could see how if I started going down that line my news editor might start – I mean the *travel* editor might start getting a bit baffled as to why I'm supposed to be writing a travel article when I've gone on a sort of eco rant and, you know, I'm there to do a job which is to try and explain to people why they should or should not go to Tasmania, not to do a forensic number on various diseases facing Tasmanian wildlife … you're trying to tell people why they might want to go there or indeed in some travel pieces I've written why they don't want to go there, and to share that experience. And the problem is you get so diverted by riffing on all sorts of side issues it can get just sort-of dull for the reader and you lose your perspective of what you're supposed to be focusing on. (Doward, 2009, pers. comm., 26 August)

Reluctance to cover contemporary environmental conflict in their articles about Tasmania was not confined to journalists who had received financial assistance from the government tourism office. This suggests that professional acculturation and genre protocols are factors that can function independently of sponsorship to influence a travel journalist's decision about whether to report environmental problems. It also allows for the possibility that more individual understandings play a part in that decision-making. For example, John Flinn (2005) – travel editor of the *San Francisco Chronicle* when he visited Tasmania at his newspaper's expense – chose to avoid the forestry debate in his published article despite hearing about it from one of his sources. Nevertheless, he was prepared to mention the devil facial tumor disease. Here, however, the motivation was not so much cosmopolitan concern for the devils but an appreciation that their vulnerability gave healthy devils scarcity value: 'That was more along the lines of, "Hey, if you want to see these things don't count on them always being there" – you know, that there's something going on' (Flinn, 2009, pers. comm., 1 March). Such reasoning draws attention to the alignment of travel media and tourism industry interests rather than the potential for travel journalism to contribute to imagined cosmopolitan communities. As following sections will demonstrate, however, the implications of this for the public sphere may be more complex and nuanced than is often assumed.

Media branding and cosmopolitan concern

Despite a symbiotic relationship between the global tourism industry and international travel media, travel journalists' descriptions and advice must be reliable if they are to be of practical use to actual travelers. One way for individuals to test a publication's reliability is to take the trips they read about, but this overlooks the role played by travel journalism in helping people choose between destinations. The number of travel journalism articles that can be consumed by a publication's readers far exceeds the number of destinations they are likely to visit. In travel media, as in tourism, readers and potential readers use brands to assess the value they can expect to derive from a product, both for practical purposes and for the construction of their own identity. Travel sections in broadsheet and tabloid newspapers are generally subsumed under the brand of the parent publication, but travel magazines must forge their own reputations. Glossy high-end travel publications often include claims of specialist knowledge and integrity when positioning their products in the market. For example, *National Geographic Traveler* employs the slogan 'Nobody Knows This World Better' (National Geographic Traveler, 2012); *Travel + Leisure* describes itself as 'the *authority* for the discerning traveller' (Novogrod in Travel + Leisure, 2012, original emphasis); and *Condé Nast Traveler* promises 'Truth in Travel' (Condé Nast Traveler, 2012).

Advertising from the tourism industry is a strong incentive for newspapers and magazines to remain on good terms with government tourism offices and tourism operators in the destinations they cover, even if their staff and freelance journalists do not accept hosted travel. Nevertheless, a travel publication's brand may benefit from editorial content containing 'constructive' criticism of an individual destination if, from a media-branding perspective, it can be seen as contributing to a high standard of service to readers and being in the long-term best interests of the destination itself. This may be the case even when, as in most instances, criticism is generally absent from the publication's editorial content, or absent from other editorial content about the same destination.

While some travel journalists have a narrow definition of tourism, others are prepared to define it according to the entire destination's branding: if a place's brand promises an unspoiled natural environment, these journalists may feel entitled to refer to environmental conflict in their articles. During the 2000s, a number of international travel journalists published articles that praised Tasmania's environment as a world-class tourism asset but expressed concern for the future of its forests. This suggests that destinations that make bold claims of exemplary environmental stewardship invite scrutiny from publications with strong brands of their own. For Jonathan Tourtellot, founding director of National Geographic's Center for Sustainable Destinations, brand promises are non-negotiable:

> if it is a touring style tourism situation, or an R&R style situation – restaurant and recreation type tourism – the place is part of the tourism product. And so if the place has forests in it, that's part of the product. If the place is supposed to have forests but doesn't, that product has been altered. And I'm putting it in cold economic terms because that's sometimes the only way you get traction. But the industry forgets that its product is the place very often. That's beginning to change, but only recently. (Tourtellot, 2009, pers. comm., 24 October)

Tourtellot's trip to Tasmania was not funded by the government tourism office but he interacted with its staff while he was in the state

(Tourtellot, 2009, pers. comm., 24 October). His published criticisms, embedded in otherwise celebratory copy about the destination, supported his magazine's own branding while representing tourism in distant destinations as a positive force capable of bringing about social change for environmental good. In this sense, his comments were consistent with a discourse described by Hajer (1995) as ecological modernization, whereby ecotourism is endorsed as capable of generating money that contributes to the management of natural places and builds local support for conservation projects. Confirming Tasmania's right to an international reputation as 'an ecotourism paradise' but drawing attention to the 'one big "except"' of logging, Tourtellot invited his readers to extend the notion of a global community of tourism consumers to include environmental concern for distant destinations. While representing Tasmanians as caring about their forests – presenting as evidence the government's boast that 40 per cent of the island was protected – he also reported that some old-growth forests were still threatened and ended his article with an overt call to action with cosmopolitical resonances: 'Visit Tasmania, and help a logger find a job in tourism' (Tourtellot, 2006, p. 38).

Travel journalists and imagined cosmopolitan communities of risks

Some interviewees who referred to environmental conflict in their published texts had connections with tourism NGOs, some spoke in terms of the global value of the world's diminishing natural places, and some described environmental politics as a distinctive feature of their writing. Often, however, they were also motivated by an awareness of similar issues at home. And although these journalists did not call upon people to take political action that would damage the tourism industry (for example, a boycott) on behalf of Tasmania's forests, Mark Jenkins, who was a columnist for *Outside* magazine in the US when he wrote about Tasmania in 2005, felt readers who were engaged in their own environmental struggles at home would make imaginative connections and gain a sense of camaraderie from reading about Tasmania's conflicts:

> I would have to just call it more an awareness ... to recognize that those people who are fighting the good fight for old-growth timber in Oregon, they've got compatriots down on the other side of the planet who are doing the same thing, and for them to recognize that we are all together on this and that it matters – that those are worthy things to fight for. (Jenkins, 2009, pers. comm., 20 March)

Jenkins described himself as a 'global correspondent' but this did not mean he necessarily saw himself as serving a universal audience. He wrote very consciously for a North American market, but in so doing he sought to help his readers appreciate the vulnerability of the world's diminishing environmental 'gems', wherever they might be (Jenkins, 2009, pers. comm., 20 March):

> I have the same problem in my own state in Wyoming where you've got foresters who don't really get the fact that they're cutting down some of the last stands that will ever exist, because the climate's changing ... [H]aving been to the Congo, the Amazon, all over Asia, all over Africa, all over Europe, all over South America, all over North America, I recognize that there are these tiny gems left, and they're very small and there are very few of them, and I kind-of believe in trying to protect every one of them ... (Jenkins, 2009, pers. comm., 20 March)

In his published article, Jenkins highlighted the cosmopolitan significance of Tasmania's environmental conflict by drawing attention to a long-term protest maintained by people from a variety of counties. In a camp in the Styx Forest, activists supported by the Wilderness Society and Greenpeace had gained international media coverage by webcasting their five-month-long tree-sit 64 meters up in the forest canopy in the Australian summer of 2003–2004 (for information about this protest, see Lester and Hutchins, 2009). In his article, Jenkins quoted an activist who spoke of ongoing work to make the station and surrounding forest a life-changing destination for concerned visitors (Jenkins, 2005).

As a high profile outdoor adventurer and author, Jenkins had considerable cultural capital. When he worked for *Outside*, he had his own column called 'The Hard Way'. His publisher paid all his travel expenses and it was his understanding that he had complete discretion over the content of his articles (2009, pers. comm., 20 March). The following comments from freelance US travel journalist Stephen Metcalf, who published an article about Tasmania in *Travel + Leisure* in 2008, explain what appears to be a more usual situation for staff or freelance travel journalists with high cultural capital. His views are insightful for their detailed explication of how the high-end travel media business model benefits from allowing writers a degree of independence:

> I've never, ever had the open conversation with an editor of either a book review or a travel outlet in which they've said, 'Look, you can't

piss off our advertisers,' but I think that's only because we all know that, and we all start from that assumption, and I understood that I was writing what was essentially a travel piece. Now the interesting thing is [that] within that understanding there's some room to play. And one of the reasons there's some room to play is that ... people do want to read journalism and they want the journalism to be very distinct from the advertising. (Metcalf, 2009, pers. comm., 26 June)

Metcalf did not accept free transport or accommodation while he was in Tasmania but, like Flinn and Tourtellot, he had some interactions with the government tourism office (Metcalf, 2009, pers. comm., 26 June). In common with another of my interviewees (Greenwald, 2008), in his 2008 article he quoted a passionate local as a 'voice of the side effects' (Beck, 1992; see Waisbord and Peruzzotti, 2009) of potential environmental damage. And although he did not describe Tasmania's environment directly as globally significant, his own concern for it as an international traveler gave this impression, as did other observations in his text, such as his comment that its air, soil and waters were 'some of the least contaminated on the planet' (Metcalf, 2008). His article powerfully exemplifies a cosmopolitan fusing of the local and global, the different and the similar, the precious and the profligate, the cultural and the political, which my research found to be an identifiable and occasionally prominent – if rare overall – feature of some international travel journalism about Tasmania in the first decade of the 21st century. Here, Metcalf's appeal to his readers to appreciate that their lives are somehow connected with the lives of distant others takes the form of a reference to Alice Waters, a prominent United States restaurateur and noted campaigner for organic produce, followed by an explicit reminder that Tasmania is 'like every place on earth' in its vulnerability to 'the forces of exploitation' (Metcalf, 2008).

The wicked problem of climate change

When analyzing the sample of articles about Tasmania by those interviewees who mentioned its contemporary environmental problems in their texts, three similarities were strongly evident: none of the articles openly advocated reader activism on behalf of Tasmania's forest, such as tourism boycotts or letter-writing campaigns; all embedded their comments or criticisms in text that praised Tasmania, implying that the destination was worth the time-consuming and expensive journey required to visit it; and none mentioned the contribution emissions from long-haul flights made to climate change.

Tasmania's capital, Hobart, is one of the farthest destinations from London and Washington – some 17,000 kilometers from each. In 2008, the UNWTO reported that:

> [l]ong-haul travel by air between the five UNWTO world tourism regions represents only 2.2% of all tourist trips, but contributes 16% to global tourism-related CO_2 emissions ... mitigation initiatives in the tourism sector will need to strategically focus on the impact of some particular forms of tourism (i.e., particularly those connected with air travel) if substantial reductions in CO_2 emissions are to be achieved. (UNWTO and UNEP, 2008, p. 34)

Some of the interviewees who had written in their travel journalism about the impacts of climate change in vulnerable destinations in other articles (for example, Flinn, 2002) or about environmental problems in Tasmania (for example, Fair, 2000, whose visit was funded by the government tourism office) were skeptical about the validity of airline carbon offset programs or the relative contribution of long-haul flights to greenhouse gases, as indicated in the following quotes from Flinn and *BBC Wildlife Magazine* travel editor Fair:

> [G]reen travel's very trendy, and whenever anything's trendy I tend to get very suspicious. It's like carbon offsets for airplanes. I once went into about four different sites they have online where you can calculate the carbon and offset, or how much carbon you're going to produce through whatever, and I picked one particular flight, and all four sites gave me radically different information about the amount of carbon it would put into the atmosphere and how many trees it's going to take to fix it up. I mean, like an order of magnitude, to the point where it almost just seemed like they were making up numbers out of thin air. So it got me very suspicious. (Flinn, 2009, pers. comm., 1 March)

> I suspect most of the carbon emissions from flights out of Britain are probably more short-haul than actually long-haul, even though you're going a lot further. I would think the sheer numbers of people going to the continent and so on probably account for far more. (Fair, 2009, pers. comm., 17 March)

At the other end of the spectrum was freelancer Paul Miles, whose comments revealed a deep ambivalence about the cosmopolitan potential of travel journalism in an age of climate change:

I think as news of climate change becomes increasingly worrying and the damage that we're causing – through, well, not just through flying obviously, but the damage that is happening to the planet is just becoming increasingly worrying – I just find it hard to rationalize promoting especially long-haul tourism, so I'm scaling back on that actually at the moment. And just doing local stuff, or trying to get away from writing about tourism. I'm writing less about tourism. (Miles, 2009, pers. comm., 9 March)

Despite having once published a mildly ironic piece in London's *Financial Times* about tourism's contribution to climate change that ended with a toast to international ecotourism (Miles, 2005), and an article about Tasmania's forestry dispute in *Condé Nast Traveler* written after a tour of the island funded by the government tourism office (Miles, 2008), Miles was frustrated by his lack of control over the content of his travel writing for mainstream publications (2009, pers. comm., 10 March; see McGaurr, 2013). His decision to write less about distant destinations and largely confine his travel journalism to pieces about local tourism in environmentally conscious media exposes one of the enduring cosmopolitan paradoxes of the genre. For if travel journalists stop writing about distant destinations because they fear this will encourage tourism that makes an unacceptable contribution to climate change, they negate their own ability to witness and mediate distant environmental threats in ways that contribute to transnational connectedness.

Conclusion

The study presented in this chapter has demonstrated that travel journalists' coverage of distant contemporary environmental problems and conflict are influenced by a variety of considerations in addition to whether or not the journalist has traveled as the guest of a government tourism office or a tourism operator. Although relatively rare, examples exist of sponsored journalists reporting environmental conflict in some of the most prestigious and high-circulation travel publications in the western world. Conversely, travel journalists who do not accept free travel and accommodation regularly ignore highly prominent environmental conflicts. Professional acculturation, personal beliefs, in-house editorial decisions and networks of interest or concern can all play a part in determining the extent of political content in individual articles.

The results of this study also suggest that travel journalists' descriptions of very particular environmental conflicts thousands of kilometers from their home markets may, indeed, bring 'the global other' imaginatively into the midst of their readers (Beck, 2011, p. 1348). In mainstream travel journalism, cosmopolitan concern finds expression in various combinations of celebration, criticism, consumerism and conflict. When it looks beyond place branding and destination image campaigns (see MacCannell, 2011, p. 210) to reflexively unveil distant environmental threats and conflict, it contributes to the mediation of a world where 'everybody is connected and confronted with everybody', whether or not they wish to know them on the deepest intercultural level (Beck, 2011, p. 1348). Such concern may also conceivably form a bridge between disparate communities engaged in similar kinds of environmental struggle, helping create the conditions for social change. In contrast to Beck's vision of imagined communities of global risks being initiated by an awareness of shared vulnerability *independent* of compassion or normative intent (Beck, 2011), however, cosmopolitical travel journalism concerned with environmental problems tends to promote an 'ethics of care' (Szerszynski and Urry, 2002) by using the language of affect to build relationships between readers, distant environments and the other.

To varying degrees, international travel journalists who chose to include environmental conflict in their coverage of Tasmania demonstrated in their interviews a sense that conflict is part of the human condition, the Earth is a shared, finite resource, the plight of local places of global value is of interest to distant audiences, and distant audiences experiencing similar conflicts will draw a sense of community from learning of others' battles. However, although a number of interviewed travel journalists used their published texts to encourage readers to be concerned about Tasmania – on one occasion going so as far as to describe it as a 'global treasure' (Greenwald, 2008) – the cosmopolitan promise of the genre was limited on a number of fronts. For example, journalists generally did not make the connection between Tasmanian conflicts and their readers' own environmental struggles explicit, perhaps because they assumed readers would make the imaginative leap without assistance, but possibly also because they did not want to trespass into the territory of domestic travel or news journalism. More importantly, a general neglect of the issue of greenhouse gas emissions from long-haul travel in their published texts about Tasmania suggests a structural impediment to the mediation of cosmopolitan concern about the long-term environmental sustainability of the global tourism

industry. Although the genre can sometimes accommodate cosmopolitan concern for destinations it simultaneously praises as places worth
visiting, long-haul transport is still so fundamental to the consumption
of distant places that it continues to be represented as a 'good' rather
than a 'bad'.

Note

1. The occupations of the journalists mentioned in this chapter are as at the
time of their visit to the state.

References

Anderson, Alison (1997) *Media, Culture and Environment*, London: Routledge.
Anderson, Alison (2000) 'Environmental Pressure Politics and the "Risk Society"',
in Stuart Allan, Barbara Adam and Cynthia Carter (eds), *Environmental Risks
and the Media*, London: Routledge, pp. 93–104.
Beck, Ulrich (1992) *Risk Society: Towards a New Modernity*, M. Ritter (trans.),
London: Sage.
Beck, Ulrich (2006) *The Cosmopolitan Vision*, Ciaran Cronin (trans.), Cambridge:
Polity.
Beck, Ulrich (2009) *World at Risk*, Ciaran Cronin (trans.), Cambridge: Polity.
Beck, Ulrich (2011) 'Cosmopolitanism as Imagined Communities of Global Risk',
American Behavioral Scientist, 55.10, pp. 1346–61.
Beck, Ulrich and Levy, Daniel (2013) 'Cosmopolitanized Nations: Re-imagining
Collectivity in World Risk Society', *Theory, Culture & Society*, 30.2, pp. 3–31.
Calhoun, Craig (2002) 'The Class Consciousness of Frequent Travelers: Toward
a Critique of Actually Existing Cosmopolitanism', *The South Atlantic Quarterly*,
101.4, pp. 869–97.
Cannon, Hugh and Yaprak, Attila (2002) 'Will the Real-World Citizen Please
Stand Up! The Many Faces of Cosmopolitan Consumer Behavior', *Journal of
International Marketing*, 10.4, pp. 30–52.
Cocking, Ben (2009) 'Travel Journalism: Europe Imagining the Middle East',
Journalism Studies, 10.1, pp. 54–68.
Condé Nast Traveler (2012) *About*. Available at: http:www.condenast.com/
brands/conde-nast-traveler.
Daye, Marcella (2005) 'Mediating Tourism: An Analysis of the Caribbean Holiday
Experience in the UK National Press', in David Crouch, Rhona Jackson
and Felix Thompson (eds), *The Media and the Tourist Imagination*. London:
Routledge, pp. 14–26.
Doward, Jamie (2004) 'Devils and the Deep, Blue Sea', *Observer*, 18 April, 'Escape',
p. 10.
Evers Consulting Services (1984), *South West Tasmania Tourism Study: Main Report*,
Evers Consulting Services.
Fair, James (2000) 'Explorer's Guide', *BBC Wildlife Magazine*, July, pp. 84–85.

Fickling, David (2004) 'Tasmanian Boycott Urged over Threat to Forests', *guardian.co.uk*, 22 March. Available at: www.guardian.co.uk/world/2004/mar/22/animalwelfare.environment.

Flinn, John (2002) 'Bearing Witness: Are Churchill's Most Famous Residents on Thin Ice?', 22 December. Available at: http://www.sfgate.com/travel/article/BEARING-WITNESS-Are-Churchill-s-most-famous-2744033.php#page-1.

Flinn, John (2005) 'A Devil of a Time in Tasmania', *San Francisco Chronicle*, 23 January, viewed 7 January 2012. Available at: http://articles.sfgate.com/2005-01-23/travel/17357830_1_nick-mooney-tasmanian-wallabies.

Flyvbjerg, Bent (2001) *Making Social Science Matter: Why Social Inquiry Fails and How It Can Succeed Again*, Steven Sampson (trans.), Cambridge: Cambridge University Press.

Fürsich, Elfriede and Kavoori, Anandam P. (2001) 'Mapping a Critical Framework for the Study of Travel Journalism', *International Journal of Cultural Studies*, 4.2, pp. 149–71.

Geertz, Clifford (1973) *The Interpretation of Cultures*, New York: Basic Books.

Greenwald, Jeff (2008) 'Sympathy for the Devil', *Islands Magazine*, 22 April. Available at: www.islands.com/article/Sympathy-for-the-Devil.

Hajer, Maarten (1995), *The Politics of Environmental Discourse: Ecological Modernization and the Policy Process*, Oxford: Clarendon Press.

Hannerz, Ulf (2006) *Two Faces of Cosmopolitanism: Culture and Politics*, Barcelona: Fundació CIDOB.

Hanusch, Folker (2011) 'Representations of Foreign Places outside the News: An Analysis of Australian Newspaper Travel Sections', *Media International Australia*, 138, pp. 21–35.

Holton, Robert J. (2009) *Cosmopolitanisms: New Thinking and New Directions*, Basingstoke: Palgrave Macmillan.

Jenkins, Mark (2005) 'Bush Bashing', *Outside*, June, available at: www.outsideonline.com/adventure-travel/Bush-Bashing.html.

Lester, Libby and Hutchins, Brett (2009) 'Power Games: Environmental Protest, News Media and the Internet', *Media, Culture & Society*, 31.4, pp. 579–95.

MacCannell, Dean (2011) *The Ethics of Sightseeing*, Berkeley: University of California Press.

McGaurr, Lyn (2013) 'Not So Soft? Travel Journalism, Environmental Protest, Power and the Internet', in Libby Lester and Brett Hutchins (eds), *Environmental Conflict and the Media*. New York: Peter Lang, pp. 93–104.

Metcalf, Stephen (2008) 'Tasmania's Gourmet Paradise', *Travel + Leisure*, February. Available at: www.travelandleisure.com/articles/tasmanias-gourmet-paradise.

Miles, Paul (2005) 'One Long Guilt Trip', *Financial Times*, 28 May. Available at: http://www.ft.com/intl/cms/s/0/c626ea46-cf14-11d9-8cb5-00000e2511c8.html#axzz2TPCmJTc8.

Miles, Paul (2008) 'Tasmania's Forest Under Threat', *Condé Nast Traveller*, March, p. 34.

National Geographic Traveler (2012) *About Traveler Magazine*. Available at: http://travel.nationalgeographic.com/travel/traveler-magazine/about-us/.

Rubin, Herbert J. and Rubin, Irene (1995) *Qualitative Interviewing: The Art of Hearing Data*, Thousand Oaks: Sage.

Santos, Carla Almeida (2004) 'Framing Portugal: Representational Dynamics', *Annals of Tourism Research*, 31.1, pp. 122–38.

Szerszynski, Bronislaw (2006) 'Local Landscapes and Global Belonging: Toward a Situated Citizenship of the Environment', in Andrew Dobson and Derek Bell (eds), *Environmental Citizenship*. Cambridge, MA: Massachusetts Institute of Technology, pp. 75–100.

Szerszynski, Bronislaw and Urry, John (2002) 'Cultures of Cosmopolitanism', *The Sociological Review*, 50.4, pp. 461–81.

Szerszynski, Bronislaw and Urry, John (2006) 'Visuality, Mobility and the Cosmopolitan: Inhabiting the World from Afar', *The British Journal of Sociology*, 57.1, pp. 113–31.

Tourtellot, Johathan (2006) 'Greenish Tasmania', *National Geographic Traveler*, March, p. 38.

Travel + Leisure (2012) *Mission*. Available at: http://www.tlmediakit.com.

United Nations World Tourism Organization (2011) *World Tourism Organization: UNWTO*. Available at: http://dtxtq4w60xqpw.cloudfront.net/sites/all/files/docpdf/aboutunwto.pdf.

United Nations World Tourism Organization (n.d.) *Tourism and Poverty*. Available at: http://step.unwto.org/en/content/tourism-and-poverty-alleviation-1.

United Nations World Tourism Organization and United Nations Environment Programme (2008) *Climate Change and Tourism: Responding to Global Challenges*. Available at: http://sdt.unwto.org/sites/all/files/docpdf/climate2008.pdf.

Urry, John (2002) *The Tourist Gaze*, 2nd edn, London: Sage.

Urry, John, and Larsen, Jonas (2011) *The Tourist Gaze 3.0*, 3rd edn, London: Sage.

Voase, Richard (2006) 'Creating the Tourist Destination: Narrating the "Undiscovered" and the Paradox of Consumption"', in Kevin Meethan, Alison Anderson and Steve Miles (eds), *Tourism Consumption and Representation: Narratives of Place and Self*. Wallingford: CABI, pp. 284–99.

Waisbord, Silvio and Peruzzotti, Enrique (2009) 'The Environmental Story that Wasn't: Advocacy, Journalism and the Asambleísmo Movement in Argentina', *Media, Culture & Society*, 31.5, pp. 691–709.

14
The Spectacle of Past Violence: Travel Journalism and Dark Tourism

Brian Creech

Introduction

Traveling to sites that relate to disaster, tragedy and death has become an established form of tourism. The aim of this chapter is to explain the conflicted role travel journalism can play in promoting this so-called dark tourism. The question is how travel journalists – who tend to focus on more positive, light-hearted stories – produce and negotiate the boundaries, motivations and ethics of this type of macabre tourism. As a case study, this chapter investigates the ways in which US-based travel journalists have participated in the public discourses that surround Tuol Sleng, a former Cambodian primary school that became a secret prison during the Khmer Rouge era and now exists as museum. Cambodia sits on the cusp of modernity, as tourists are lured to the country by the promises of exotic beauty but also by the darkness of a violent recent history, fueling a booming tourism industry that has led to real economic gains for the country (Chheang, 2009). By examining travel journalism articles +that document visits to and histories of the site, this chapter posits that travel journalism can operate as a realm of discursive practice that helps make sense of complex realities by offering, beyond tourism's broader commercial concerns, a mode for engaging with dark sites that preserves empathy.

In the wake of decades of war and the brutal violence of the Khmer Rouge regime, tourism operators and state institutions have been able to parlay international interest in Cambodian history into a multi-billion dollar industry (Heikkila and Peycam, 2010). The Angkor temples and other sites of Khmer antiquity have been popular destinations for nearly a century, but in recent decades, the violent specters that haunt the sites of the Khmer Rouge regime have also sparked tourist interest,

particularly around one of Pol Pot's most grisly institutions: S-21, the
Tuol Sleng prison. The Khmer Rouge regime is an undeniably dark part
of Cambodian history. As the country struggles to legitimize its current
political system, the era remains a difficult topic for public discourses
to address. This chapter argues that travel journalism, while not resolving any of the tensions that circulate around dark sites, offers a style
of engagement for foreign audiences for connecting discursively with,
at best, an empathy and humanism that accounts for the interplay
between complex histories and contemporary conditions.

Dark tourism: sites and spectators

Dark tourism is a contested practice with an equally contested definition.
As the Institute for Dark Tourism Research (2013, p. 1) defines it, dark
tourism is the 'act of travel to sites, attractions and exhibitions of death,
disaster or the seemingly macabre', and more specifically, an 'often-
contentious consumer activity that can provoke debate about how death
and the dead are packaged up and consumed within the modern visitor
economy'. The definition foregrounds tourism as a consumer activity by positing dark tourism as an act of commodification with death
and macabre history as the object of consumption (Sharpley, 2009).
Furthering this understanding, Lennon and Foley (2000, p. 3) state dark
tourism indicates a 'fundamental shift' in the way that tour operators
offer products and tourists consume them. However, as Bowman and
Pezzullo (2009) argue, positing a difference between dark tourism and
more traditional and normative forms of tourism delimits dark travel
as solely a form of consumption. Doing so forecloses any analyses that
attempt to understand how aspects of dark tourism may constitute
broader cultural practices and rituals memorializing death beyond the
commercial structure of tour operators and tourists.

Sites of memorialized death and destruction can be thought of as the
physical manifestation of the discourses and practices that give expression to the anxieties and realities of war that inflect late modernity
(Foucault, 2007; Soja, 1989). Dark tourism, Lennon and Foley (2000)
argue, is a distinctly postmodern practice that allows for simultaneous
stabilization, exploitation, commodification and sanctification of these
sites, while history, politics, geography and culture provide important,
meaning-making context. More so, these abstractions are given physical form via the memorialization of dark sites (Portegies et al., 2011).
These sites constitute a locus between a victimized population's need for
historical reconciliation and that same population's economic need for

the income that tourism brings, particularly in the case of countries like Cambodia (Edkins, 2005; Beech, 2009). Overt commodification, understood as 'kitschification', exists as a normative limit to this touristic memory work, providing the means for critique when the production of certain buildings with dark histories as tourist sites has gone 'too far' (Potts, 2012, p. 247).

By investigating the relationship between dark tourist sites and travel journalism about those sites, this chapter acknowledges what Sharpley and Stone (2009) conclude in their work – namely, that broader historical forces may give shape to the contested meaning surrounding these sites, but the site itself mediates a complex relationship between the tourist and the broader culture. These sites are understood through ideological and discursive regimes, often as a corollary to less macabre mainstream culture (Dale and Robinson, 2011). For instance, Levitt (2012) offers that Hollywood death tours exist as attendant aspects of the broader practices of celebrity culture, while Pezzullo (2009) finds that tourism surrounding Hurricane Katrina elicits a politically charged remembrance aimed at implicating the Bush administration and the broader American political apparatus in the death and destruction associated with the storm. The multitudinous meanings of dark tourist sites may give rise to a counter-hegemonic understanding of reality, with dark sites acting as a collection of artifacts granting emblematic form to contemporary anxieties over the potential for cultural, political or material destruction (Stone, 2013).

While this research indicates that dark tourism is far more complex than mere commercialization of death and destruction, commodification remains an uneasy concern for many scholars and commentators (Burmon, 2010; Osbaldiston and Petray, 2011). Tourists emerge as individuals implicated in the commercialization of dark sites, as several applied research studies of tourist behavior overtly seek ways to translate these tourists' motivations into profit, using surveys, focus groups and in-depth interviews to evaluate effective commodification strategies around dark sites (Biran, Poira and Oren, 2011; Di Giangirolamo, 2012; Kang et al., 2011; Podoshen, 2013). This chapter does not attempt to alleviate the tension and concern over the commercialization of dark sites. Instead, it argues, using travel journalism on Cambodia's Tuol Sleng as a case study, that commodification and commercialization is merely an aspect of the tourist experience. Dark tourism remains a complicated practice connected to the cultural and economic vitality of many sites, and any singular form of journalism can neither resolve nor catalog the varied manifestations of these tensions. However, as a mode of meaning-making, travel journalism offers texts that not only curate dark sites while also providing relevant context and history, but also render sensible the possible emotional reactions and broader cultural forces a viewer may experience. Despite tensions between commodification and the search for authentic experience, travel journalism offers an arena for public discourse where the relationship between sites, spectators and broader cultural and historical forces plays out.

Theoretically situating travel journalism and dark tourism

A key theoretical question underpins the study of travel journalism and dark tourism: How might scholars and observers make sense of these stories, videos and images in a way that does not reduce them to mere representation, but also takes into account the complicated politics and cultural issues surrounding stories about dark places? Dark tourism in the global South is a quintessential postmodern practice, bound up with the contemporary politics of globalization by repackaging international atrocities for a style of Western travel consumption that simultaneously legitimates and exploits victims while also offering a vector for the movement of capital from the first to the third world. As Chouliaraki (2010) explains, the contemporary representations of third-world suffering often operate amid a field of 'post-humanitarianism', (p. 121) where the sympathy and action they elicit keep tacit broader critiques of the social, economic and political forces that allow that suffering to persist.

In order to address the overt question of post-colonial politics embedded in the practice of dark tourism, and journalism about dark tourism, it is important to avoid questions about what journalism should do and instead attend to what it already does. While normative concerns may offer ethical guideposts for practitioners, they do not explain the complex situation of the bodies of those who suffered through these crises (Isaac and Ashworth, 2012). As Spivak (1988) reminds us, postcolonial bodies are bound up in many-layered discursive regimes, inscribed by relations of power that are neither monolithic nor singular in their origins. Casper and Moore (2009) carry this notion into the modes of public knowledge production that make bodies sensible and offer an analysis of how victimhood and sympathy become broadly understood discursive categories with material consequences. Their approach fits with the vision of travel journalism forwarded by this chapter: as a cultural form, travel journalism is implicated in the practices that help American and Western audiences make sense of foreign atrocities.

As a cultural and discursive form, dark travel journalism offers a specific rationality, emotional grammar or sympathetic language for approaching and understanding the truly abhorrent. To argue for a reflexive understanding of travel journalism is to argue that, at its best, the field offers a means to interrogate the material and ineffable traces of history, and amid this interrogation, continue the never-ending project of making sense of the impersonal forces that drive modernity: capitalism, globalization, industrialization, war and catastrophe. As Sontag (2003, p. 106) has argued:

> The argument that modern life consists of a diet of horrors by which we are corrupted and to which we gradually become habituated is a founding idea of the critique of modernity – the critique being almost as old as modernity itself.

While the dubious morality of viewing suffering in any capacity may seem apparent, the question is in fact much more complicated. Representation and atrocity are tightly bound, as the 'expectation of photographic evidence' betrays an underlying logic of witnessing (Sontag, 2003, p. 83). Nothing can be made sense of until it is seen, and it is through subsequent acts of synthesis and analysis that understanding begins to emerge.

In this regard, we can begin to understand how travel journalism about dark places fits within a broad continuum of witnessing divorced from a temporally based need to witness the act itself. As Allan (2013, p. 100) notes, the act of witnessing does not need to be considered solely an act rooted in the empirical authority of an individual's eyes and ears, it is also 'a discursive act, where one's own experience is stated for an audience elsewhere'. Journalists, as witnesses, become part of a situated act, inscribed by cultural, social, political, professional and geographic forces that inform their abilities to witness and represent their experience. In a strictly Foucauldian sense (Foucault, 1972), the travel journalist occupies a negotiated subjectivity. From amid the field of discourse that is travel journalism, the journalist has the particular capability to experience dark places, consider their history and represent the broader historical and cultural issues at play to a global audience. Through the very act of visiting a dark site and then choosing to construct a narrative about it, the travel journalist creates a singular document, which will have a broader effect on the field of knowledge surrounding a particular site by creating a narrative about that site, its history, or the phenomenological experience of the visit for an audience (Foucault, 2001).

Still, the fraught nature of dark tourism and travel journalism should remain sharply in the foreground. Zelizer's (2010) analysis of news images of subjects on the verge of death offers a helpful example in dealing with the particular consequences of specific images. Though images of dead individuals offer a vast vocabulary for making sense of the abstract concept of death, no image, story or representation offers a complete understanding. Instead, there is an imaginative, subjunctive capacity to these depictions that 'underscores a paradoxical relationship with death's visibility' (Zelizer, 2010, p. 319). In short, when journalists attempt to represent dark sites and the motivations that compel visitors, they operate in a variety of capacities: as evidence gatherers, guides, explainers and analyzers. Each of these roles gains its power through the imaginative capacity elicited in the audience, as these texts, in their specificity, continue the arduous task of making sense of atrocity.

Case study: Cambodia's Tuol Sleng prison

Tuol Sleng, or S-21, offers an important site for scholarly analysis into the relationship between dark tourism and travel journalism. Over the past decade, tourism has provided an economic boon to a country previously ravished by decades of civil war (Tyner, 2008). As a site and a text, Tuol Sleng distills the violence of the Khmer Rouge genocide into a memorial attracting nearly 30 per cent of all Cambodian tourists each year (Chheang, 2009). In order to understand how Tuol Sleng operates discursively within the realm of travel journalism, we briefly must turn to Cambodian history to understand how the practices of discourse, as a discontinuous set of tactics and activities rooted in history, have constructed the site.

History of Tuol Sleng and the Khmer Rouge genocide

In 1975, after the Khmer Rouge forced Phnom Penh's residents out of the city and into rural work camps, a former math teacher and Khmer Rouge commander named Duch took over Tuol Sleng, remaking the former primary school into a prison that specialized in torture (Chandler, 1999). Between 1975 and 1979, an estimated 17,000 people were held, tortured and killed inside the former primary school. The Vietnamese Army liberated Phnom Penh in 1979 and by January 1980 bureaucrats turned Tuol Sleng into a memorial and museum modeled after holocaust memorials like Dachau (Hinton 1998).

Officials at Tuol Sleng kept meticulous records of every prisoner that had come through, including photos, typed personal histories, notes on tortures and elicited confessions. The prison's sole purpose was to ferret out dissent within Democratic Kampuchea through any means necessary, as the soldiers and prison staff used torture to elicit forced confessions from prisoners (Brinkley, 2011). Guards would shock prisoners with car batteries, hang them by their arms and beat prisoners in order to force them to implicate neighbors and family members in plots to overthrow Angkar – the Khmer word for organization that became the name for the Khmer Rouge leaders. Eventually, the lens of scrutiny turned inward as the paranoia spread and members of the Khmer Rouge leadership cadre began to accuse each other of intended sedition against Angkar (Ly, 2003). Chandler (1999, p. 155) argued that this spread of the torture and punishment apparatus at Tuol Sleng reflects 'tendencies toward acculturation and obedience' and that a 'paranoid logic at the heart of institutionalized operations was emboldened by a systematic isolation from the rest of the world'.

Other historical analyses of Tuol Sleng root it within the material workings of the Khmer Rouge, and in doing so, attempt to avoid characterizing the regime as irrational and anomalous, instead arguing that they were driven by a singular, internally coherent ideology. Artifacts and sites like Tuol Sleng stand as material evidence of this ideology. Ben Kiernan's (2002) work attempts to place Cambodia's past into a global political context, notably criticizing the role of US intervention and military activity in Southeast Asia through the 1960s and 1970s as the primary factor that destabilized the Cambodian monarchy, polarized the Cambodian population against its government, fostered extremism, and created a power vacuum as US state department officials supported Pol Pot politically and militarily as he tried to assume power. In this analytical scheme, Tuol Sleng becomes an unforeseen consequence of shortsighted foreign policy driven by contention between capitalist and Communist ideological battles.

Elsewhere, Kiernan has also criticized the ways that the US's, Soviet Union's, and China's international military and diplomatic machinations set the stage for the violence and genocide of Democratic Kampuchea. Given that, according to Kiernan (2004, p. 76), the Khmer Rouge genocide is not taught in textbooks or discussed in Cambodian classrooms or the Cambodian press, the country and its citizens have suffered from a particular historical amnesia, or at least dysmorphia, where 'half a millennium of intermittent civil conflict, foreign invasions and even genocide not only devastated Cambodia, but also prevented the Khmer people from weighing their experiences in historical perspective'. It is amid this murk of history that Tuol Sleng stands as a material referent for historical violence whose consequences have been difficult to fully conceive and render in collective memory (Winter, 2008).

Travel journalism as contemporary context

In order to read Tuol Sleng as something more than a collection of artifacts, one must encounter the space armed with previous knowledge. This knowledge allows the viewer to interpret the space as physical justification for the prior interpretations that they have encountered, precisely because Tuol Sleng attempts to preserve a sense of material accuracy, embodying the history of Cambodian violence as it occurred (Williams, 2004). The artifacts and the buildings, though, exist as little more than the setting of a historical story, and so others have filled the narrative void. Travel writers, politicians and survivors have all told stories about the space, and in this process of story-telling, have offered an interpretation of the history embodied there (Burmon, 2010; Taing, 2012). In this respect, Tuol Sleng's sparseness allows it to exist as an open text that can accommodate various interpretations of Cambodia and its recent history.

Much of the American and British travel journalism about Tuol Sleng recounts the four years of Khmer Rouge rule and contextualizes the site's significance within that history. Such articles create an informative backdrop grounded in the materiality of the prison. These histories also perform another important task: demarcating the relevant events within an era that no longer holds sway over the present, yet may be mined for lessons by intuitive observers. For example, Ben Ehrenreich's (2009) account of Tuol Sleng and the nearby killing fields treats the skulls, artifacts and buildings as relics from a different time, powerless in their existence but potent in the messages they reveal about the potential for human violence. The travel journalist, as a visitor to the site, is uniquely configured to articulate the relevant history and lessons that may be learned. Ehrenreich, like other travel journalists before him, wanders through the halls of Tuol Sleng, recounting what he sees – the artifacts, torture implements and prisoner photos – while also artfully weaving relevant history into his observations. From here, he synthesizes the site, the relevant history and contemporary Cambodian politics, and the complexity of tourism into a journalistic text. The article reveals an informed analysis of the site that goes beyond the immediate reality to discern a broader cultural concern:

A small, dark-skinned man dusted the display cases and the edges of the picture frames with a rag. He did not dust the skulls in front of me, but they, too, were very clean. Whether his or someone else's, it was clearly someone's job to dust the skulls. Most of the Cambodians I spoke with objected strongly to this exhibition, and you don't have to know much about Buddhism to understand why people would not want bones that might belong to their loved ones displayed for the entertainment of foreign tourists. (Ehrenreich, 2009, p. 66)

Such a passage is indicative of the potential truths that careful travel journalism may reveal about a dark tourist site. In the hands of certain journalists these sites become the means by which Westerners might begin to discern a complicated and dark history without overtly commodifying death and destruction.

Other articles take on a less analytic and more traditional guidebook style, but still reveal the cognitive effect that such spaces should have on a viewer. These are articles composed with the individual tourist as the intended audience, but also reveal a normative position of cosmopolitan universalism in statements such as:

Altogether, a visit to Tuol Sleng is a profoundly depressing experience. The sheer ordinariness of the place makes it even more horrific: the suburban setting, the plain school buildings, the grassy playing area where children kick around balls juxtaposed with rusted beds, instruments of torture and wall after wall of disturbing portraits. It demonstrates the darkest side of the human spirit that lurks within us all. Tuol Sleng is not for the squeamish. (Lonely Planet, 2012, p. 1)

Such passages invoke an ideal spectator, informed of international histories and capable of empathetically understanding what he or she views. Still, such articles reveal an individuating orientation implicit in the production of dark tourist sites, providing discursive instruction on the proper way to consume these sites: 'Visitors walking through the hallways of this former-high-school-turned-prison must confront the pain, uncertainty and fear of thousands of victims looking back at them from the black-and-white photographs taken by prison guards' (Sites, 2006, p. 1).

Anthony Bourdain's print and television journalism about Cambodia and its history serve as exemplary texts, displaying the individuating power of travel discourse and the ideal that a tourist experience, especially a dark tourist experience, should affect the individual viewer intellectually and emotionally. In an episode of No Reservations dedicated to Cambodia, Bourdain (2011) laments his inability to understand the country, its history and its dark sites on prior trips, often stating throughout the program that he hopes that he has the maturity and clarity of purpose to encounter the country and its sites as an empathetic individual. Such statements fit within Bourdain's (2008) assertion that travel writing reveals to the reader the landscape of an inner journey more than an outer journey. In this regard, travel journalism offers the audience discursive tools for properly consuming a travel experience in a way that is at least aware of the broader context within which the traveler is embedded. Experience becomes reflexive, impacting the individual psyche and consciousness.

Still, such moves keep the individual traveler as the primary locus of tourism, relegating all experience to the individual's capacity to render such experiences as meaningful. Take, for instance, a guide to Cambodia featured in Conde Nast Traveler (2012, p. 34), where Tuol Sleng is listed as an attraction among other tourist options. Describing the site alongside restaurants and spas, as a 'grisly reminder of Cambodia's tragic past', the magazine maintains a consumer-focused perspective when advising, 'a visit is essential to understanding how far the country has progressed'. As one New York Times article points out, consumption-oriented travel has provided a necessary economic boon to the country, explicating a 'luxury revolution' in a way that distills the complex interplay of bloody history and commodified modernity with the statement that 'change has come at an amazing pace' (Gross, 2006, p. 22).

At its extremes, traditional travel journalism renders meaningful or affective individual experiences as an expected part of a package tour that capitalizes on the imbued importance of history. Khmer Rouge history and violence serve as the backdrop of a larger trip, as travel writers praise the relatively cheap luxuries and pleasures available to savvy tourists (Alford, 2009; Cohane, 2012). Such articles are evidence of a discourse that overtly celebrates tourism as an economic boon for the Cambodian economy, as reflected in Burmon's (2010, p. 1) travel observations: 'In 2008, the road between Anlong Veng and Angkor Wat was widened and paved, allowing easy access for tour buses. Near the old house of Ta Mok, a wooden sign tacked to a tree reads: TOURISM WILL BRING MONEY AND JOBS.'

Though the economic value of Tuol Sleng as a site leads some observers to lament impending commodification of sacred sites, most travel journalism surrounding Tuol Sleng negotiates the apparent conflict between the economic pressures of tourism with the historical reverence the site deserves. By looking towards Cambodia's future and the economic promise that tourism portends, journalists like Hari Kunzru (2007) articulate a historical narrative rooted in the atrocity of past violence but find optimism in the way that tour operators turn that history into successful business ventures. Take for example, the following passage:

> Today's Phnom Penh has come a long way from the haunted, empty place of 1975, when the Khmer Rouge drove its entire population out into the countryside to grow rice. It's a pleasant city, with ... an ease and friendliness that will no doubt soon make it one of the most popular destinations in Asia. (Kunzru, 2007, p. 18)

By placing a violent history among a contemporary context, travel journalists offer observations of a current moment, distilling the forces of history and economy into the seeming presence of opportunity, albeit an opportunity contingent upon international travelers.

Such accounts hinge upon the careful use of details to append an emotional register onto the experience. In the following passage, Kunzru (2007) finds an object in the prisoner photos that line many of the rooms in Tuol Sleng, which gives form to the phenomenological experience of walking through the museum, placing them in context with the rest of the museum's space:

> Tuol Sleng is almost unbearable. Not because of the classrooms partitioned by crudely built brick walls into tiny cells. Not even because of the display of shackles and torture instruments or the lurid paintings done by one of the survivors. The hardest part is seeing the faces of the victims. Everyone brought to S-21 had their picture taken, numbers round their necks, clamped into a device to keep their heads still for the camera's shutter. There are rooms of 10-by-eights of dead people, men, women and children, even tiny babies, 'discarded' (in the jargon of the interrogators) because of their perceived threat to the paranoid members of the Central Committee. (Kunzru, 2007, p. 17)

These uses of observable details anchor the affective tenor of the experience into a concrete reality, providing for the audience a corollary for the experience, should they visit the same site. In the language of sensory experience and observed detail, Istvan (2003, p. 16) attempts to bridge the gap between Cambodia's history and present:

Today the place looks benign, with palm trees and grass lawns in a suburban setting. From the outside, Tuol Sleng could be a school anywhere in the world. But inside are weapons of torture, skulls, blood stains and photographs of thousands of people who were murdered.

The contrasting details stand of their own accord, a synecdoche for more ineffable forces that escape journalistic expression.

Yet the ideal international traveler may deploy a keen sense of understanding that at least attempts to carefully understand and interpret the travel experience, as typified by the careful prose of Ian Buruma, whose writing on Cambodia was anthologized in America's Best Travel Writing 2008. Buruma's (2007, p. 2) writing combines typical travel suggestions and restaurant and hotel reviews with recent history, politics, personal narrative and cultural observation, all distilled into sentences that portray a broader theme: 'A typical Southeast Asian city, then. And yet, there is a melancholy about the place that still speaks of recent horrors. Phnom Penh is a city of survivors.' Buruma's (2007, p. 4) text further creates an experiential distance when he states in reflexive prose, 'despite all the photos and grisly exhibits, the mind is unable to conceive what it must have been like to have been a prisoner in Tuol Sleng'. This observant distance allows Buruma (2007, p. 5) to occupy an idealist discursive position capable of authoritatively stating, 'as at many of these atrocity sites, tourists go in giggling, or chatting about last night's barhopping adventures ... But by the end of their time there, silence prevails. Even the unimaginable can be difficult to bear'.

Such understanding and nuance, while laudable, may not always be achievable. As Spano (2011, p. 22) notes of her own travels, 'my aspirations were modest: to learn, if not to understand, how the unthinkable happened in Cambodia'. Journalists like Spano, who admit to the limitations of their own understanding, still curate experiences for the reader, at the very least drawing attention to important sites, the relevant history and political context that make them important. As economic development continues to churn forward in Cambodia and more overtly commercial forms of tourism become the norm, Spano (2011, p. 23) notes how visitors may pass through important landmarks 'without understanding the significance of bullet marks'. As an engaged observer, Spano models a form of reflexive tourism, culminating in the final paragraph:

> Standing at Pol Pot's grave, I mentally retraced the road I'd taken
> through Cambodia, showing how all the conditions were present

that had allowed the Khmer Rouge to take power: poverty, igno-
rance, misgovernment, radical ideology, foreign intervention. The
only additional element needed was the psychopath buried at my
feet. (Spano, 2011, p. 23)

As an observer injected into the prose, Spano occupies the role of a different kind of travel guide, showing readers the cognitive routes they
may follow to help make sense of the experience, remaining ever aware
of the scars of history while also trying to make sense of the contemporary moment. Though she distills all of Cambodia's problematic history
into the persona of Pol Pot, Spano offers a means for making sense of
contemporary Cambodia through journalistic personalization, a narrative tool that locates complex historical events within the biographies,
bodies and decisions of individual figures.

More generally, travel journalists evoke a sense of critical distance,
aware of the relevant history, but evoking a perspective untouched by
it. It is this distance that lends them authority, as if their travelogues are
the blank canvases on which history, culture and politics are mixed with
contemporary experience. A version of deeper 'truth' emerges within
the text itself, the product of a skilled interlocutor and synthesizer. As a
collection of texts, the travel journalism surrounding dark sites like Tuol
Sleng does not offer any overt means of resolving the litany of apparent tensions that constitute the site itself. Yet, as the above examples
reveal, certain examples of travel journalism offer a mode of engagement that situates the site as a material symbol where an awareness of
the economic, political, cultural and historical forces that inscribe dark
sites may emerge. The travel journalist takes up a particular position,
capable of fashioning details into a resonant narrative that takes into
consideration previously unseen and underlying complications, as the
following passage from Iyer (1999, p. 34), tying Tuol Sleng to the trial
of the prison's former warden, Duch shows:

For 20 years now, Tuol Sleng has been a notorious memorial to the
Khmer Rouge killers who ruled Cambodia from 1975 to 1979. ...
Display cases are littered with the hoes and shovels and iron staves
they used to brain people to death; along the walls, hundreds upon
hundreds of black-and-white faces stare back at you, dazed or ter-
rified, recalling the people, often children and often themselves
Khmer Rouge executioners, who were executed here. ... But this
spring the monument to the past came into the news again when
the man who had overseen the torture for four years, Kang Khek

Ieu, generally known as Duch, was suddenly discovered, by foreign journalists, in a western Cambodian village. ... Yet every prospect of new sunlight in Cambodia brings shadows, and justice itself seems a rusty chain that will only bloody anyone who tries to touch it. To try the Khmer Rouge chieftains would be, in a sense, to prosecute the whole country: almost everyone around – from the exiled King Sihanouk to the one-eyed Prime Minister to the man next door – has some connection to the Khmer Rouge killers.

In its most ideal state, travel journalism about dark sites strikes a broad perspective and an explanation, attaching complicated, abstract relations of history and culture to key, specific details. As a cultural product emblematic of a style of tourism, it at the very least offers a discursive position capable of distilling the darker forces that underscore contemporary reality in a way that is not overdetermined by the commercial pressures of the tourism industry.

Conclusion

Undergirding the analysis of dark tourist sites, like Tuol Sleng, is the assumption that history leaves its mark on the things it touches. As the preceding analysis shows, individual sites begin to bear the marks of multiple discourses, tracing over time the workings of power in a tangible and observable way. Public discourses, like travel journalism, offer a means for making sense of the ways these multiple discourses may structure a contemporary tourist experience (Prior, 2011).

As Hughes (2008) has argued, tourists seeking sites like Tuol Sleng often come to an understanding of a destination's culture as symptomatic of a single, dark historic event, casting a sinister pallor over the history that follows. In dealing with dark sites though, travel journalism can offer a narrative for understanding how these sites are produced and what they mean. In order to overcome the crassly commercial implications of certain kinds of tourism, travel journalists occupy an idealist discursive position removed from the immediate circumstances and economic opportunities of dark sites. Their narration notes contingent and conflicted histories and cultures and aims at eliciting genuine empathy while negotiating apparent tensions between pleasure and enlightenment implicit in travel to dark sites. For the broader study of travel journalism, this means taking the journalists seriously not just as the chronicler of experiences, but as discursive guides, lending audiences the affective and cognitive tools for making sense of the broader world. This is an act very much complicit in the larger forces of globalization. But by treating each tourist site and act of travel journalism as a singular object of inquiry, scholars may begin to understand the various modes of meaning-making travel journalists deploy and, in turn, invite audiences to use as their own tools.

References

Alford, Henry (2009) 'Banishing Ghosts in Cambodia', *New York Times*, 12 March, Available at: http://travel.nytimes.com/2009/03/15/travel/15cambodia. html?pagewanted=all

Allan, Stuart (2013) *Citizen Witnessing: Key Concepts in Journalism*, Malden, MA: Polity Press.

Beech, John (2009) 'Genocide Tourism', in Richard Sharpley and Phillip R. Stone (eds), *The Darker Side of Travel: The Theory and Practice of Dark Tourism*. Buffalo: Channel View Publications, pp. 207–23.

Biran, Avital, Poria, Yaniv and Oren, Gila (2011) 'Sought Experiences at (Dark) Heritage Sites', *Annals of Tourism Research*, 38.3, pp. 820–41.

Bourdain, Anthony (2008) 'Introduction', in Anthony Bourdain (ed.), *America's Best Travel Writing 2008*. New York: Houghton Mifflin, pp. 2–12.

Bourdain, Anthony (2011) 'Cambodia', Anthony Bourdain: No Reservations. [Television Broadcast]. The Travel Channel. Available at: http://www.travel channel.com/tv-shows/anthony-bourdain/episodes/cambodia

Bowman, Michael S. and Pezzullo, Phaedra C. (2009) 'What's so "Dark" about "Dark Tourism"?: Death Tours and Performance', *Tourist Studies*, 9.3, pp. 187–202.

Brinkley, Joel (2011) *Cambodia's Curse*, Philadelphia: Perseus.

Burmon, Andrew (2010) 'Dark Tourism', *The Atlantic Monthly*, November. Available at: http://www.theatlantic.com/magazine/archive/2010/11/dark-tourism/8250/

Buruma, Ian (2007) 'Phnom Penh Now', *Travel + Leisure*, April. Available at: http://www.travelandleisure.com/articles/phnom-penh-now

Casper, Monica J. and Moore, Lisa Jean (2009) *Missing Bodies: The Politics of Visibility*, New York: New York University Press.

Chandler, David (1999) *Voices from S-21*, Berkeley: University of California Press.

Chheang, Vannarith (2009) 'State and tourism planning: A case study of Cambodia', *Tourismos*, 4.1, pp. 63–82.

Chouliaraki, Lilie (2010) 'Humanitarian Communication Beyond a Politics of Pity', *International Journal of Cultural Studies*, 13.2, pp. 107–26.

Cohane, Ondine (2012) 'Cambodia's Sweet Spot', *New York Times*, 2 March, Available at: http://www.nytimes.com/2012/03/04/travel/cambodia-in-and-around-kep-open-but-undeveloped.html?pagewanted=all

Conde Nast Traveler (2012) 'Guide to Phnom Penh', Available at: http://www.cntraveller.com/guides/asia/cambodia/phnom-penh/where-to-stay

Dale, Crispin and Robinson, Neil (2011) 'Dark Tourism', in Peter Robinson, Sine Heitman, and Peter Dieke (eds), *Research Themes for Tourism*. London: CAB International Publishing.

Di Giangirolamo, Gianluigi (2012) 'Dark Tourism and Horror's Travel: Some Italian Tourist Sites', *AlmaTourism: Journal of Tourism, Culture, and Territorial Development*, 3.5, pp. 123–5.

Edkins, Jenny (2005) 'Exposed Singularity', *Journal for Cultural Research*, 9.4, pp. 359–86.

Ehrenreich, Ben (2009) 'Cambodia's Wandering Dead: The Ghosts of Genocide Pay Penance for Western Guilt', *Harper's*, April, pp. 59–66.

Foucault, Michel (1972 [1967]) *The Archeology of Knowledge*, A.M. Sheridan Smith (trans.), New York: Pantheon.

Foucault, Michel (2001) *Fearless Speech*, Joseph Pearson (ed.), London: Semiotext(e).

Foucault, Michel (2007[1964]) 'The Language of Space', in Jeremy Crampton and Stuart Elden (eds), *Space, Knowledge and Power: Foucault and Geography*. Burlington: Ashgate, pp. 164–8.

Gross, Matt (2006) 'Why is Everyone Going to Cambodia?' *New York Times*, 22 January, Available at: http://travel2.nytimes.com/2006/01/22/travel/22cambodia.html?pagewanted=1&_r=0

Heikkila, Eric J. and Peycam, Philippe (2010) 'Economic Development in the Shadow of Angkor Wat: Meaning, Legitimation and Myth', *Journal of Planning Education and Research*, 29.3, pp. 294–309.

Hinton, Alexander Laban (1998) 'Why Did You Kill?: The Cambodian Genocide and the Dark Side of Face and Honor', *Journal of Asian Studies*, 57.1, pp. 93–122.

Hughes, Rachel (2008) 'Dutiful Tourism: Encountering the Cambodian Genocide', *Asia Pacific Viewpoint*, 49.3, pp. 318–30.

Institute for Dark Tourism Research (2013) 'Research at the IDTR'. Available at: http://www.dark-tourism.org.uk/research

Isaac, Rami K. and Ashworth, Gregory J. (2012) 'Moving From Pilgrimage to "Dark" Tourism: Leveraging Tourism in Palestine', *Tourism, Culture, & Communication*, 11.3, pp. 149–64.

Istvan, Zoltan (2003) '"Killing Fields" Lure Tourists in Cambodia', *National Geographic*, 10 January, Available at: http://news.nationalgeographic.com/news/2003/01/0110_030110_tvcambodia.html.

Iyer, Pico (1999) 'Into the Shadows', *Time Magazine*, 16 August. Available at: http://content.time.com/time/world/article/0,8599,2054340,00.html

Kang, Eun-Jang, Scott, Noel, Lee, Timothy and Ballantyne, Roy (2011) 'Benefits of Visiting a "Dark Tourism" Site: The Case of the Jeju April 3rd Peace Park, Korea', *Tourism Management*, 33.2, pp. 257–65.

Kiernan, Ben (2002) *The Pol Pot Regime: Race, Power and Genocide in Cambodia Under the Khmer Rouge, 1975–1979*. New Haven: Yale University Press.

Kiernan, Ben (2004) 'Recovering History and Justice in Cambodia', *Comparitiv*, 14, pp. 76–85.

Kunzru, Hari (2007) 'A New Day Dawns', *The Guardian*, 12 May, Available at: http://www.guardian.co.uk/travel/2007/may/13/escape.generalfiction

Lennon, John and Foley, Malcolm (2000) *Dark Tourism: The Attraction of Death and Disaster*, London: Continuum.

Levitt, Linda (2012) 'Solemnity and Celebration: Dark Tourism Experiences at Hollywood Forever Cemetery', *Journal of Unconventional Parks, Tourism & Recreation Research* 4.1, pp. 20–25.

Lonely Planet (2012) 'Review of Tuol Sleng Genocide Museum', Available at: http://www.lonelyplanet.com/cambodia/phnom-penh/sights/museum/tuol-sleng-museum

Ly, Boreth (2003) 'Devastated Vision(s): The Khmer Rouge Scopic Regime in Cambodia', *Art Journal*, 62.1, pp. 66–81.

Osbaldiston, Nick and Petray, Theresa (2011) 'The Role of Horror and Dread in the Sacred Experience', *Tourist Studies*, 11.2, pp. 175–90.

Pezzullo, Phaedra C. (2009) 'Tourist and/as disasters: Rebuilding, Remembering, and Responsibility in New Orleans', *Tourist Studies*, 9.1, pp. 23–41.

Podoshen, Jeffrey S. (2013) 'Dark Tourism Motivations: Simulation, Emotional Contagion, and Topographic Comparison', *Tourism Management*, 35, pp. 263–71.

Portegies, Ariane; Haan, Theo D.; Issac, Rami; and Roovers, Lucette (2011) 'Understanding Cambodian Tourism Development through Contextual Education', *Tourism Culture & Communication* 11.2, pp. 103–116.

Potts, Tracey J. (2012) '"Dark Tourism" and the "Kitschification" of 9/11', *Tourist Studies*, 12.3, pp. 232–49.

Prior, Nick (2011) 'Postmodern Restructurings', in Sharon MacDonald (ed.), *A Companion to Museum Studies*. Malden: Blackwell, pp. 509–24.

Sharpley, Richard (2009) 'Dark Tourism and Political Ideology: Towards a Governance Model', in Richard Sharpley and Phillip R. Stone (eds), *The Darker Side of Travel: The Theory and Practice of Dark Tourism*. Buffalo: Channel View Publications, pp. 145–66.

Sharpley, Richard and Stone, Phillip R. (2009) 'Life, Death, and Dark Tourism: Future Research Directions and Concluding Comments', in Richard Sharpley and Phillip R. Stone (eds), *The Darker Side of Travel: The Theory and Practice of Dark Tourism*. Buffalo: Channel View Publications, pp. 247–51.

Sites, Kevin (2006) 'Torture Chamber: The Tuol Sleng Museum Shows the Atrocities of Khmer Rouge', *Yahoo News*, 24 July, Available at: http://www.redding.com/news/2006/jul/24/torture-chamber-the-tuol-sleng-museum-shows-the/

Sontag, Susan (2003) *Regarding the Pain of Others*. New York: Picador.

Spano, Susan (2011) 'Cambodia after the Killing Fields', *Los Angeles Times*, 15 May, Available at: http://www.latimes.com/travel/la-tr-killing-fields-20110515,0,425351.story?page=1\

Spivak, Gayatri (1988) 'Can the Subaltern Speak?' in Patrick Williams and Laura Chrisman (eds), *Colonial Discourse and Post-Colonial Theory: A Reader*. New York: Columbia University Press, pp. 66–111.

Stone, Phillip R. (2013) 'Dark Tourism, Heterotopias and Postapocalyptic Places: The Case of Chernobyl', in Leanne White and Elspeth Frew (eds), *Dark Tourism and Place Identity: Managing and Interpreting Dark Places*. New York: Routledge, pp. 88–102.

Soja, Edward (1989) *Postmodern Geographies: The Reassertion of Space in Critical Social Theory*, New York: Verso.

Taing, Judy (2012) 'Cambodia: The Kingdom that Cries Wolf', *Article 19*, 17 November, Available at: http://www.article19.org/resources.php/resource/3533/en/cambodia:-the-kingdom-that-cries-wolf

Tyner, Justin (2008) *The Killing of Cambodia: Geography, Genocide and the Unmaking of Space*, Burlington, VT: Ashgate.

Williams, Paul (2004) 'Witnessing Genocide: Vigilance and Remembrance at Tuol Sleng and Choeung Ek', *Holocaust and Genocide Studies*, 18.2, pp. 234–54.

Winter, Tim (2008) 'Postconflict Heritage and Tourism in Cambodia: The Burden of Angkor', *International Journal of Heritage Studies*, 14.6, pp. 524–39.

Zelizer, Barbie (2010) *About to Die: How News Images Move the Public*, New York: Oxford University Press.

Name Index

Subject Index

Printed and bound by CPI Group (UK) Ltd, Croydon, CR0 4YY